Springer Proceedings in Complexity

More information about this series at http://www.springer.com/series/11637

Matthäus P. Zylka • Hauke Fuehres
Andrea Fronzetti Colladon • Peter A. Gloor
Editors

Designing Networks for Innovation and Improvisation

Proceedings of the 6th International COINs Conference

 Springer

Editors
Matthäus P. Zylka
Department of Information Systems
 & Social Networks
University of Bamberg
Bamberg, Germany

Hauke Fuehres
Department of Information Systems
 & Social Networks
University of Bamberg
Bamberg, Germany

Andrea Fronzetti Colladon
Department of Enterprise Engineering
Tor Vergata University of Rome
Rome, Italy

Peter A. Gloor
MIT Center for Collective Intelligence
Cambridge, MA, USA

ISSN 2213-8684   ISSN 2213-8692   (electronic)
Springer Proceedings in Complexity
ISBN 978-3-319-82639-4   ISBN 978-3-319-42697-6   (eBook)
DOI 10.1007/978-3-319-42697-6

Printed on acid-free paper

This Springer imprint is published by Springer Nature
The registered company is Springer International Publishing AG
The registered company address is: Gewerbestrasse 11, 6330 Cham, Switzerland

Preface

This book contains 17 peer-reviewed contributions presented at the Sixth international COINs Conference, held in Rome, Italy, from 8 to 11 June 2016.

The papers in this book cover a broad range of topics, starting with an analysis of different communities and societies through social network analysis (SNA). Classic SNA looks at the structure of networks; the papers in this book add analysis of the dynamics of network change over time, and an analysis of the content of the networks, for example, in e-mails, Tweets, or Wikipedia. Dynamic and content-based SNA affords an X-ray into the inner workings of an organization, mapping the informal relationships that transcend organizational hierarchy. It gives an assessment of communication and knowledge flow, resulting in actionable data to optimize outcomes. The approach pursued in these papers puts a lens to the organization by mining e-mail archives and, as relevant, other electronic communications from online social media (e.g., Twitter, Facebook, Wikipedia, Reddit, and other online forums) to make existing communication patterns visible. A second group of papers looks at design patterns and pattern language for creativity and other business processes, such as design patterns for education, creating workshops or modeling collaboration in the kitchen. This book is divided in four parts taking the aforementioned wide range of research fields into account. The four parts correspond to the paper sessions held at the conference.

The first part of this book contains four papers about communities, societies, and culture. Robin Gieck, Hanna-Mari Kinnunen, Yuanyuan Li, Mohsen Moghaddam, Franziska Pradel, Peter A. Gloor, Maria Paasivaara, and Matthäus P. Zylka talk about "Cultural Differences in the Understanding of History on Wikipedia." Leanne Ma, in her paper "The Emergence of Rotating Leadership for Idea Improvement in a Grade 1 Knowledge Building Community" investigates the role of rotating leadership in the classroom. Iroha Ogo, Satomi Oi, Jei-Hee Hong, and Takashi Iba present a method for reconsidering strengths of a community in their study "Creating Community Language for Collaborative Innovation Community." Takashi Iba concludes the first part of this book with his study titled "Sociological Perspective of the Creative Society."

The second part contains four papers about machine learning, prediction, and networks. Veikko Isotalo, Petteri Saari, Maria Paasivaara, Anton Steineker, and Peter A. Gloor talk about "Predicting 2016 US Presidential Election Polls with Online and Media Variables," whereas Johannes Bachhuber, Kim Rejstrom, Christian Koppeel, Jeronim Morina, and David Steinschulte present a different approach analyzing the US Presidential Elections in their work titled "US Election Prediction—A Linguistic Analysis of US Twitter Users." In the study "Only say something when you have something to say"—Identifying Creatives Through Their Communication Patterns" by Peter Gloor, Hauke Fuehres, and Kai Fischbach, the authors study the communication patterns of particularly creative people in the R&D department of a global energy firm through their e-mail communication. Finally, Matteo Cinelli, Giovanna Ferraro, and Antonio Iovanella in their study "Some Insights into the Relevance of Nodes' Characteristics in Complex Network Structures" present a methodology that can be used as a pre-processing tool for avoiding the inclusion of non-effective nodes' characteristics.

The third part contains four papers about design patterns. This part starts with "Patterns as a Supporting Tool for Workshop Generators" by Yuma Akado, Masafumi Nagai, Taichi Isaku, and Takashi Iba. Next, Norihiko Kimura, Hitomi Shimizu, Iroha Ogo, Shuichiro Ando, and Takashi Iba present "Design Patterns for Creative Educational Programs." The next study by Takashi Iba, Ayaka Yoshikawa, Norihiko Kimura, Tomoki Kaneko, and Tetsuro Kubota, "Pattern Objects: Making Patterns Visible in Daily Life," proposes the concept of pattern objects to make contents of pattern languages visible in daily life. The third part concludes with the study by Taichi Isaku and Takashi Iba, "From Chefs to Kitchen Captains: A Leader Figure for Collaborative Networks in the Kitchen."

The final part contains five papers about social media and social networks. Peter Gloor, Andrea Fronzetti Colladon, Christine Miller, and Romina Pellegrini start this part with their study titled "Measuring the Level of Global Awareness on Social Media," where they introduce a novel approach to measure the degree of global awareness by analyzing social media. Subsequently, Joao Marcos de Oliveira and Peter A. Gloor present an application that extracts newsworthy user-generated content from Wikipedia and Twitter in "The Citizen IS the Journalist—Automatically Extracting News from the Swarm." Next, Timo Herttua, Elisa Jakob, Sabrina Nave, Rambabu Gupta, and Matthäus P. Zylka explore the essence, definition and methods of the grassroots practitioner term growth hacking in their study called "Growth Hacking: Exploring the Meaning of an Internet-born Digital Marketing Buzzword." Then, action-development-relationship (ADR) processes as a social innovation-design methodology for creating strategic partnerships and networks are presented by Makoto Okada, Yoichiro Igarashi, Hirokazu Harada, Masahiko Shoji, Takehito Tokuda, and Takashi Iba in "ADR Processes for Creating Strategic Network for Social Issues: Dementia Projects." The book concludes with the study of Sayaka Sugimoto titled "Depression as a Global Challenge and Online Communities as an Alternative Venue to Develop Patients-led Supportive Network."

We wish to express our gratitude to Agostino La Bella who delivered the inaugural keynote about leadership, communication, and charisma, as well as to the other two invited keynote speakers Jana Diesner (Words and Networks: Using Natural Language Processing to Enhance Graphs and Test Network Theories), and Peter A. Gloor (Building Collective Consciousness—Homo Collaborensis). Both, Peter A. Gloor and Jana Diesner conducted also workshops at the conference. In total, four workshops were held. We also thank the other two workshop instructors, Takashi Iba (Design Patterns of Creativity Workshop), and Lukas Zenk (Designing Innovative Networking Events Workshop), for insightful and creative workshops.

We are pleased to acknowledge the important help of the colleagues who assisted in the organization of this event, starting with Agostino La Bella (Conference Chair), Andrea Fronzetti Colladon (Local Chair), and the staff members at the Tor Vergata University of Rome, without whom the conference could not have been organized.

Further, we would like to thank the *Steering Committee*, responsible for the development and support of the COINs conference series, whose members are Peter A. Gloor (MIT), Ken Riopelle (Wayne State Univ.), Julia Gluesing (Wayne State Univ.), Takashi Iba (Keio Univ.), Casper Lassenius (Aalto Univ.), Maria Paasivaara (Aalto Univ.), Christine Miller (IIT), Cristobal Garcia (Pontificia Univ. Católica de Chile), and Andrea Fronzetti Colladon (Tor Vergata Univ. of Rome).

The conference was supported by many educational and organizational sponsors: Department of Enterprise Engineering at the Tor Vergata University of Rome, Italian National Research Council (CNR), Associazione dei Laureati in Ingegneria di Tor Vergata (Alitur), Italian Association of Business Engineering (AiIG), MIT Center for Collective Intelligence, Wayne State University, Aalto University, and the Pontificia Universidad Católica de Chile. On behalf of all the participants, we would like to thank those supporters.

Finally, we would like to thank all the authors and reviewers for their contributions to this book.

Bamberg, Germany   Matthäus P. Zylka
Bamberg, Germany   Hauke Fuehres
Rome, Italy   Andrea Fronzetti Colladon
Cambridge, MA, USA   Peter A. Gloor

Contents

Part I
Communities, Societies and Culture

Chapter 1
Cultural Differences in the Understanding of History on Wikipedia

Robin Gieck, Hanna-Mari Kinnunen, Yuanyuan Li, Mohsen Moghaddam, Franziska Pradel, Peter A. Gloor, Maria Paasivaara, and Matthäus P. Zylka

1.1 Introduction and Related Work

In August 2015, Wikipedia has 280 (active) different language editions facilitating in multiple communication channels between different languages and cultures. Previous research validated Wikipedia as a great data provider for resolving different research questions (Medelyan et al. 2009; Schroeder and Taylor 2015; Xu and Li 2015). Most studies regarding differences between various language editions focused on the interaction and communication of editors with only a few papers aimed to investigate cultural similarities and dissimilarities.

The MIT Media Lab developed the project Pantheon that used Wikipedia data and information of Murray's "Human Accomplishment" (2003) to map cultural production (Yu et al. 2015). Cultural differences in Wikipedia have been examined by analyzing editorial behavior in various Wikipedia editions (e.g., Nemoto and Gloor 2011) or by investigating the description of cultural practices in Wikipedia like food cultures (Laufer et al. 2015). Other studies focused on important historical

R. Gieck
University of Bamberg, Bamberg, Germany

H.-M. Kinnunen • M. Paasivaara
Aalto University, Helsinki, Finland

Y. Li • M. Moghaddam • F. Pradel
University of Cologne, Cologne, Germany

P.A. Gloor
MIT Center for Collective Intelligence, Cambridge, MA, USA
e-mail: pgloor@mit.edu

M.P. Zylka (✉)
Department of Information Systems and Social Networks,
University of Bamberg, Bamberg, Germany
e-mail: matthaeus.zylka@uni-bamberg.de

© Springer International Publishing Switzerland 2016
M.P. Zylka et al. (eds.), *Designing Networks for Innovation and Improvisation*,
Springer Proceedings in Complexity, DOI 10.1007/978-3-319-42697-6_1

persons by using network analysis (Aragon et al. 2012; Eom and Shepelyansky 2013; Gloor et al. 2015), quantitative and qualitative content analysis (e.g., Callahan and Herring 2011), or different kinds of discussion spaces (Hara et al. 2010). Furthermore, some studies focused on articles in different languages to explore controversial topics displaying the emergence of different preferences and interests in Wikipedia (Yasseri et al. 2014; Bilic and Bulian 2014).

Our paper aims to study the cultural similarities and differences between Chinese, English, German, French, and Swedish Wikipedia by focusing on articles relating to the most important and influential historical war events written in these five languages on Wikipedia. We use network, content, sentiment, and complexity analysis to investigate the cultural differences. We also have included Finnish and Arabic Wikipedia in two of our analyses. Culture can be viewed from different theoretical perspectives, for example a communicative view according to Hall (1959) "Culture is communication and communication is culture". A collaborative and shared-knowledge platform like Wikipedia combines communication and interaction between many users. Since Wikipedia articles are written in different languages, we assume that the editors of the articles are influenced by their own culture (e.g., Nemoto and Gloor 2011). First of all the English and Chinese speaking population consists of people from different cultures so when we are talking about the English, Chinese, or even German culture we classify them more as Culture clusters according to House et al. (2004). For example, the English speaking nations are combined in an "Anglo cluster" consisting of Canada, USA, Australia, and England. There is also a Chinese speaking cluster called "Confucian Cluster" including China, Singapore, Hong Kong, etc., and a Germanic cluster of German speakers in Austria, Germany and parts of Switzerland, and the Netherlands. Therefore, the different understandings of historical events are unavoidable. According to Laufer et al. (2015), we use Wikipedia language as the proxy for cultural communities and also according to Yu et al. (2015) it is known that defining transmitted information as a culture is a common practice among scholars. It is important to investigate cultural similarities and dissimilarities because different understandings of historical events can lead to conflicts and confusion.

1.2   Methods

We used articles of historical war-related events in Chinese, English, French, German, and Swedish Wikipedia which are categorized as military historical events to investigate cultural differences and similarities between languages. As Wikipedia does not contain a universal categorization for wars, we had to find another way of extracting the most important war events. Fortunately, many countries have their own article in Wikipedia that contain a list of wars in which the respective country has taken part. In this research, the historical war event is a war that one of the listed countries has participated.

In order to get the most important war-related events, we created a Java program that fetched a list of war-related events from Wikipedia, counted the indegree (number of incoming links) of each event and ordered events by their popularity. We considered the measured indegree as a key figure of importance which is in line with Charles Murray's work (2003), who used the frequency of mentioning as determination for importance. We used Wikipedia pages containing a list of wars involving the respective country for mining war events. Those pages existed both in English and the original language (Chinese, German, Swedish, and French) independently from each other. This way of data extraction enabled us to gather information of 93 English, 28 Swedish, 253 German, 201 Chinese, 69 Finnish, and 104 French war articles.

We have also included Finnish and Arabic Wikipedia analysis in some parts of our research. Finnish was skipped from further analysis because the sentiment analysis tool did not support it and Arabic was excluded because it was really hard to define war events from that area.

1.3   Results

1.3.1   Mining Cross-Cultural Relations

We used the *Jaccard similarity coefficient* to measure the similarities between the 10, 20, and 50 most important historical events (according to Laufer et al. 2015). Here, we have used this measure to compare the size of intersection for the same historical wars divided by the size of potential intersection (sample union). For example J between Chinese and English culture:

$$J = (\text{Chinese} \cap \text{English}) / (\text{Chinese} \cup \text{English}).$$

Table 1.1 shows the similarity among the top 20 most important events between the different languages on Wikipedia. For example, English and Chinese Wikipedia have 15 % of the 20 most important events in common (equivalent to 3 of the top 20 wars). From this result, we can conclude that the French and German culture, the German and English culture, and also the French and English culture are most similar. Chinese and Swedish cultures have the least in common with the other cultures.

Table 1.1 Similarities among the 20 most important historical wars

	Chinese	English	German	French	Swedish
Chinese	1	0.15	0.15	0.1	0
English		1	0.35	0.5	0
German			1	0.45	0.1
French				1	0.1
Swedish					1

Table 1.2 Wikipedia war-like word analysis

Word	English	German	French	Swedish	Finnish	Chinese	Arabic
Kill	0.0572	0.0236	0.0210	0.000847	0.0540	0.0510	0.575
War	190	0.0722	0.196	0.0103	0.0326	0.0853	0.148
Battle	149	0.0312	0.0686	0.00661	0.0534	0.0339	0.0719
Murder	0.0548	0.0149	0.0195	0.00510	0.0195	0.0153	0.0575
Victory	0.0650	0.0417	0.0644	0.00303	0.0497	0.0370	0.0485
Defeat	0.0390	0.0215	0.0139	0.00109	0.0200	0.0300	0.0161
Revolution	0.0539	0.0257	0.0805	0.00259	0.0145	0.0492	0.0593
Victory/defeat	1.67	1.94	4.63	2.78	2.49	1.23	3.01

1.3.2 War-Like Words Analysis

In order to investigate the popularity of war-related words, we performed an analysis conducting a search by Google to determine usage frequency of different war- and violence-related words. Words like "kill", "war", "battle", "murder", "victory", "defeat", and "revolution" were translated and used in the target language. The number of hits was divided by the size of the relevant language edition (Wikipedia 2014) to obtain comparable results.

Results listed in Table 1.2 show that English Wikipedia contains a significant amount of mentioned words such as "war" and "battle" compared to other languages while "Revolution" was named most in French Wikipedia. The ratio between victory and defeat shows how the different Wikipedias focus on victories in comparison to defeats. Based on this ratio, French Wikipedia clearly used the most positive language while Chinese was the most critical. All five Wikipedia languages focus more on victories than defeats.

1.3.3 Wikipedia Date Page Analysis

As a separate Wikipedia analysis method, we created a Java program that used the Wikimedia API for searching and analyzing the amount of war events in date pages. All our target languages have a page that contains a list of events that have happened on that day throughout history. We wanted to analyze how many of the events that are listed in those date pages were war-related compared to the total amount of events shown.

Figure 1.1 contains the results of a date page analysis divided into monthly values. It seems that Finns and Chinese focus most on war-related events in Wikipedia date pages while the French mention the least amount of war events compared to other countries. Furthermore, there are less war events during wintertime than in summer or autumn. This emphasizes that most of the war events seem to have occurred between June and October.

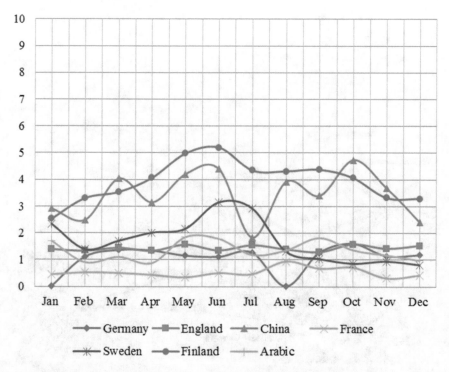

Fig. 1.1 Wikipedia date page analysis

1.3.4   Network of War Events

Based on previous results, we have been able to create a "Network of War Events" for the previously named nations plus Finland (see Fig. 1.2). For creation and analysis of the network, we used the open source tool Gephi (Bastian et al. 2009). We have developed a model of different node types and weighted edges in a directed network. As node types we defined two categories. The first category represents the considered nations (China, England, Finland, France, Germany, and Sweden) and because it is a directed network each "nation-node" has an outdegree-value of 20 and an indegree-value of 0. For a better illustration, nodes with the type "nation" are shown in different colors while nodes of the second category (type "events") are displayed in a gradient from light to dark red depending on the indegree, the importance (weighted indegree), and the number of incoming links is represented by an increasing weighting and thickness of the edges. Analysis of the graph shows that World War I with the highest Indegree (5) and Centrality (1) is the most frequently mentioned event among our analyzed nations, while World War II is the most important event by consideration of the number of incoming links. In particular, the English and German Wikipedia contribute a large amount of backlinks to World War II but this is also relative to the size of the Wikipedia Language versions.

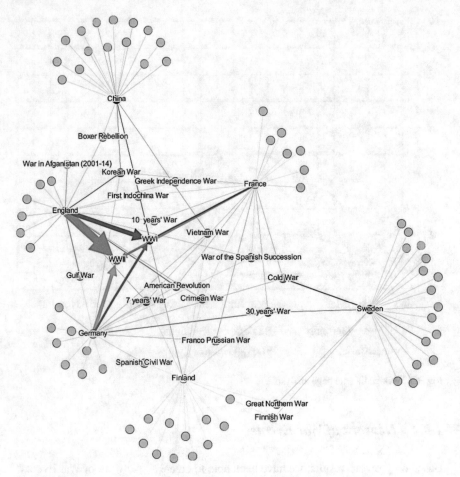

Fig. 1.2 Network of war events

1.3.5 Sentiment Analysis

Sentiment Analysis focuses on the analysis of people's sentiments, opinions, attitudes, and emotions toward elements like themes, individuals, and organizations (Serrano-Guerrero et al. 2015). We used "Semantria for Excel" to analyze the cultural differences among Chinese, English, German, French, and Swedish Wikipedia. Semantria is a multilingual sentiment engine which masters several languages such as English, German, Chinese or French and weights each paragraph of a document based on the document's components (themes, topics, entities) and their sentiment values (Semantria 2012).

Figure 1.3 shows the proportion of negative, neutral, and positive paragraphs for each language. Considering the whole content of the viewed events for each language, the Swedish Wikipedia comprises the largest proportion of negative para-

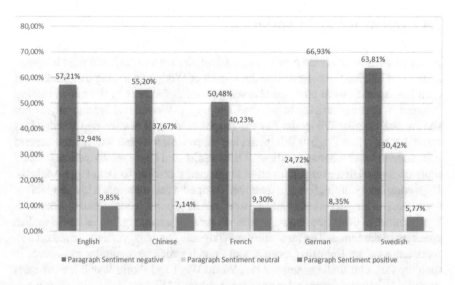

Fig. 1.3 Proportional view of the sentiment analysis for each language

graphs with 63.81 %, followed by English (57.21 %), Chinese (55.20 %), and French (50.48 %). The German Wikipedia consists of neutral evaluations to a large extent (66.93 %) and a similarity among all five languages is the low percentage of positive paragraphs. Considering the category, these results are as expected.

1.3.6   Language Complexity Analysis

We tried to determine readability, the ease with which a reader can understand a written text (Dale and Chall 1949). Gunning Fog Index (DuBay 2004) has been used to analyze the top 20 war events' content for English, French, German, and Swedish Wikipedia while because of the completely different syllable structure of Chinese compared to European languages Yang's Index, specially developed for Chinese has been used for Chinese Wikipedia (Sung et al. 2013). Due to the different interpretation of output value for these two indexes, we translated the outputs into their individual equivalent reading level to enable comparison between war-related articles written in Chinese and other languages. The most complex article of all languages is regarding the Cold War in English Wikipedia with the highest Gunning Fog index of 32.02, with the most "comfortable" Gunning Fox index being 8. The average reading level needed for Chinese war-related articles is only 16 years of age which indicates high school students in China should perceive these articles as relatively readable. On the other hand, English and Swedish high school students may perceive the language of their Wikipedia as more complex because their average reading level for this material requires an age over 25 years.

1.4   Discussion and Conclusion

In this chapter, we analyzed how war-related articles are written in different language editions of Wikipedia to determine the impact of Wikipedia among the speakers of each language. Several analyses showed general differences in the understanding between Wikipedia articles in selected languages. War-related word analysis has shown that English Wikipedia has the largest amount of war-related words. This result suggests that English Wikipedia has more violence- and war-related content compared to other language editions. The Swedish and Chinese Wikipedias focused more on war-related events than other languages according to Data Page Analysis. High complexity in English and Swedish Wikipedia and therefore a low readability could cause a long-term effect and a loss of reader groups. It could lead to a decreasing influence of these two language editions for all ages among English and Swedish speakers in the future. The most similar Wikipedia language versions based on network analysis are English, German, and French which have similar top events in their top 20s. Our findings suggest that World War I and World War II are the most important historical events relating to these cultures. Wikipedia already offers a collaboration of editors for example, who can discuss certain improvements in the "talk page" of an article. Offering additionally the option to discuss the understanding of articles in an international environment could enhance the collaborative nature of Wikipedia. One idea could be to improve the transparency about the editors' social background to identify potentially influences in the understanding of an event. In a forum, where the understanding of certain aspects can be discussed from editors and readers from different countries, could help to get an overview over different interpretations of an event.

The presented work can be improved in several ways. The extraction of historical events can be improved by finding a more systematical way of data extraction for example by using keywords in the title (e.g., "War", "Revolution", "Battle"). Furthermore, it would be interesting to analyze more language versions and cluster them by using the gathered information. According to Murray (2003), we used the number of incoming links (indegree) as measure for the importance of an event but other measures could also be used and compared with each other (e.g., number of edits or views). Another supplement could be to add a variable which weights the size of each Wikipedia language version for transnational contemplation. The event network could be improved by creating a tool that visualizes the gathered data and automatically combines it with more information, like time of the event or the people involved. Also adding more events and thus more nodes could show a higher connection rate between different Wikipedia language editions. Due to the different standard of output value in the complexity analysis, a comparison of the readability between Chinese and the other four languages is not accurate enough in this work. A generally recognized index for both Asian and European languages should be developed for future work to gain more accurate results. Finally, due to the missing sentiment scores for the Swedish language we had to perform the sentiment index analysis without these data. The development of a comprehensive tool would help to enhance the analysis without restrictions.

Despite these limitations, our analyses show that Wikipedia reflects cultural differences. We hope they will spur additional research in other settings using alternative ways of data extraction and analyses of cultural dissimilarities. While Wikipedia is increasingly significant for our everyday life, our data show the importance for further investigation of cultural dissimilarities.

References

Aragon P, Laniado D, Kaltenbrunner A, Volkovich Y (2012) Biographical social networks on Wikipedia—a cross-cultural study of links that made history. In: WikiSym'12, ACM, New York, pp 3–6

Bastian M, Heymann S, Jacomy M (2009) Gephi: An open source software for exploring and manipulating networks. In: International AAAI Conference on Weblogs and Social Media, Association for the Advancement of Artificial Intelligence, Palo Alto

Bilic P, Bulian L (2014) Lost in translation: Contexts, computing, disputing on Wikipedia. In: iConference 2014 Proceedings, University of Illinois at Urbana-Champaign, Champaign, pp 32–44

Callahan ES, Herring SC (2011) Cultural bias in Wikipedia content on famous persons. J Am Soc Inf Sci Technol 62(10):1899–1915

Dale E, Chall JS (1949) Techniques for selecting and writing readable materials. Elementary English 26(5):250–258

DuBay WH (2004) The principles of readability. Impact Information, Costa Mesa, CA

Eom YH, Shepelyansky DL (2013) Highlighting entanglement of cultures via ranking of multilingual Wikipedia articles. PLoS One 8(10), e74554

Gloor PA, Marcos J, de Boer PM, Fuehres H, Lo W, Nemoto K (2015) Cultural anthropology through the lens of Wikipedia: Historical leader networks, gender bias, and news-based sentiment. arXiv preprint arXiv:1508.00055

Hall ET (1959) The Silent Language. Doubleday, New York, p 191

Hara N, Shachaf P, Hew KF (2010) Cross-cultural analysis of the Wikipedia community. J Am Soc Inf Sci Technol 61(10):2097–2108

House RJ, Hanges PJ, Javidan M, Dorfman PW, Gupta V (eds) (2004) Culture, leadership, and organizations: The GLOBE study of 62 societies. Sage, Thousand Oaks

Laufer P, Wagner C, Flöck F, Strohmaier M (2015) Mining Cross-cultural relations from Wikipedia—A study of 31 European food cultures. In: Proceedings of the ACM Web Science Conference WebSci '15, ACM, New York, pp 1–10

Medelyan O, Milne D, Legg C, Witten IH (2009) Mining meaning from Wikipedia. Int J Hum Comput Stud 67(9):716–754

Murray C (2003) Human accomplishment: The pursuit of excellence in the arts and sciences, 800 BC to 1950. Harper Collins, New York

Nemoto K, Gloor PA (2011) Analyzing Cultural Differences in Collaborative Innovation Networks by Analyzing Editing Behavior in Different-language Wikipedias. Procedia Soc Behav Sci 26:180–190

Schroeder R, Taylor L (2015) Big data and Wikipedia research: social science knowledge across disciplinary divides. Inf Commun Soc 18(9):1039–1056

Semantria (2012) Frequently asked questions, API version 2.0. Semantria, Boston

Serrano-Guerrero J, Olivas JA, Romero FP, Herrera-Viedma E (2015) Sentiment analysis: a review and comparative analysis of web services. Inform Sci 311:18–38

Sung HT, Chen RL, Lee YX, Cha RS, Tseng HQ, Lin WJ, Chang DX, Chang GE (2013) Investigating Chinese text readability linguistic features, modeling, and validation. Chin J Psychol 55(1):75–106

Wikipedia Tilastot (2014) https://fi.wikipedia.org/wiki/Wikipedia:Tilastot. Accessed Nov 2015
Xu B, Li D (2015) An empirical study of the motivations for content contribution and community
    participation in Wikipedia. Inf Manage 52(3):275–286
Yasseri T, Spoerri A, Graham M, Kertész J (2014) The most controversial topics in Wikipedia: A
    multilingual and geographical analysis. In: Fichman P, Hara N (eds) Global Wikipedia:
    International and Cross-Cultural Issues in Online Collaboration, Scarecrow Press, Lanham
Yu AZ, Ronen S, Hu K, Lu T, & Hidalgo, CA (2015) Pantheon: A Dataset for the Study of Global
    Cultural Production. arXiv preprint arXiv:1502.07310

Chapter 2
The Emergence of Rotating Leadership for Idea Improvement in a Grade 1 Knowledge Building Community

Leanne Ma

2.1 Introduction

Creative, collaborative engagement with ideas defines Knowledge Building/knowledge creation and is fundamental to work in today's knowledge society: "At its best, the Knowledge Society involves all members of the community in knowledge creation and utilization" (United Nations 2005, p. 141). Collaborative Innovation Networks—in knowledge-creating organizations, research laboratories, and design firms—are advancing the frontiers of the knowledge society (Gloor 2005). Correspondingly, there is growing recognition that schools must shift from preparing students to be consumers to producers of knowledge (e.g., OECD 2015; Tan and Tan 2014). Knowledge Building pedagogy (Scardamalia and Bereiter 2014) represents a longstanding effort aimed at transforming education into a knowledge-creating enterprise by empowering all students to take collective responsibility for creating and advancing knowledge for public good.

Creative work requires flexible, distributed social configurations that support collaborative improvisation and group flow (Sawyer 2015). In Collaborative Innovation Networks (COINs), members have a strong sense of autonomy and self-organize around shared goals, with members rotating leadership as new tasks emerge (Gloor and Cooper 2007). Similarly, in the Knowledge Building classroom, both the teacher and the students share a sense of *collective responsibility* and all members improvise within the context of 12 Knowledge Building principles that support self-organization around idea improvement (Scardamalia 2002). For example, the principles of *idea diversity*, *improvable ideas*, and *rise above* prioritize ideas at the center of class discussions and highlight the iterative nature of idea generation, refinement, and invention in knowledge creation processes that enhance

L. Ma (✉)
Ontario Institute for Studies in Education, University of Toronto, Toronto, ON, Canada
e-mail: leanne.ma@mail.utoronto.ca

© Springer International Publishing Switzerland 2016
M.P. Zylka et al. (eds.), *Designing Networks for Innovation and Improvisation*,
Springer Proceedings in Complexity, DOI 10.1007/978-3-319-42697-6_2

the breadth and depth of group understanding and achievement. In a recent study that aimed to assess the Knowledge Building principle of *collective responsibility* (Ma et al. 2016), rotating leadership was uncovered as an emergent phenomenon in three successful Knowledge Building classes, with children as young as 6 and 9 years of age. The current study is exploratory in nature, with the goal of extending this work on rotating leadership by assessing idea improvement and addressing issues surrounding conceptual depth and coherence of student ideas. Two research questions were developed for analysis at the group and individual levels:

1. Over the course of their Knowledge Building, how many students emerge as leaders? What are the pivotal points in the discussion that indicate idea improvement and knowledge advancement?
2. What is a student doing in a leadership position? How does assuming leadership relate to learning outcomes, as reflected in student portfolios?

2.2   Methods

The study took place at the Dr. Eric Jackman Institute of Child Study in Toronto, Canada, where Knowledge Building pedagogy and technology have been implemented and refined over decades of research. The sample consists of 22 students engaged in inquiry about water, while receiving age-appropriate pedagogical and technological supports. Knowledge Forum (Scardamalia 2004) served as the central online space for students to work with ideas and build community knowledge. They contributed their ideas as notes in conceptual spaces called views; developed ideas with "build-on" notes; and synthesized ideas with "rise-above" notes. As Knowledge Forum was integrated into daily classroom practices, students engaged in continuous reading, writing, and revising of notes to advance their community knowledge. Over the course of 3 months, students wrote a total of 198 notes across three views: Water, Evaporation, and Where water goes from our houses. Student-generated problems of understanding included: the water cycle; clouds and weather systems; oceans, lakes, and rivers; sewage and waste management systems; and water as source of life. At the end of their inquiry, a fourth view was created for students to create portfolios and reflect on their understanding of the water cycle.

The student discourse in Knowledge Forum was exported into Knowledge Building Discourse Explorer (KBDeX) (Oshima et al. 2012) in order to perform content-based social network analyses. A list of content-related words (100 words) extracted from the Ontario Curriculum of Science and Technology (2007) was used as "key" ideas in the community knowledge.

KBDeX (Fig. 2.1) is an analytic tool that produces network visualizations of words, notes, and students based on the co-occurrence of keywords in each network: the learners network (top right) shows idea sharing across learners, the notes network (bottom left) shows idea overlap across notes, and the word network

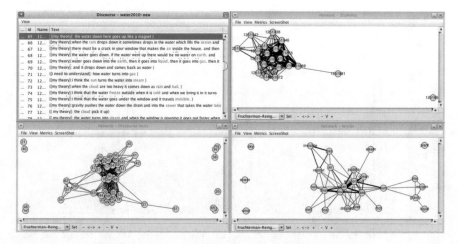

Fig. 2.1 KBDeX visualization of student discourse in Knowledge Forum

(bottom right) shows how ideas are connected in the community knowledge as group discussions progress. KBDeX also produces temporal visualizations of network metrics, such as betweenness centrality, which indicates the extent to which a member influences other members of the group, and centralization of betweenness centrality, which indicates the extent to which the network is centralized. Within a Collaborative Innovation Network, oscillating patterns of betweenness centrality (i.e., rotating leadership) is considered a good measure of group productivity and creativity (e.g., Kidane and Gloor 2007).

2.3   Findings

In this section, group level analyses (temporal network analyses, discourse analysis) are reported to address the first set of research questions, followed by individual level analyses (social network analyses, content analysis) to address the second set of research questions.

2.3.1   Overview of Knowledge Forum Activities

Descriptive measures of online behaviors in Knowledge Forum indicate that students were working productively with ideas over the course of their inquiry. On average, each student wrote nine notes (range=3–20), and used 8.7 key words (range=3–15). The most common ideas discussed by students include: "cloud", "evaporation", "rain", "sewer", "air", "earth", "ocean", "lake", "snow", and "sun".

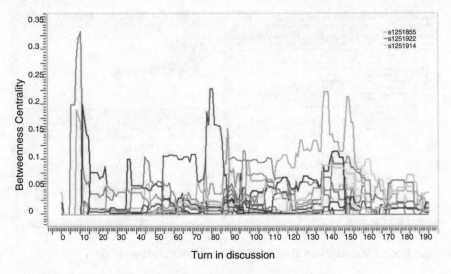

Fig. 2.2 KBDeX visualization of individual betweenness centralities across time

2.3.2 Temporal Network Analyses

The average centralization of betweenness centrality was 0.087 (range=0–0.32), indicating that the network was relatively decentralized over time. Figure 2.2 shows temporal analysis of betweenness centrality for each student in the class. The Y axis of the chart shows the betweenness centrality value, and the X axis shows the turn in discussion over time. Each colored line represents a student, resulting in the display of 22 lines in the chart. The oscillation of colored lines depicts the phenomenon of rotating leadership, which means that the leading student (i.e., the student with the highest betweenness centrality) changed frequently. Of the 22 students, 13 students took a leading position, suggesting that many students were influential at different times. The topmost influential leaders were students s1251855 (c_b=0.356), s1251922 (c_b=0.325), and s1251914 (c_b=0.324). While student s1251855 had the greatest influence in the student network, student s1251914 had the longest duration of influence in the student network.

2.3.3 Discourse Analysis

The student discourse in Knowledge Forum was arranged in chronological order to assess the collective trajectory of idea improvement and knowledge advancement. Each student note was coded along the dimensions of: (1) epistemic complexity (from unelaborated facts, elaborated facts, unelaborated explanations, to elaborated explanations) and (2) scientific sophistication (from pre-scientific, hybrid, basically scientific, to scientific). Whereas epistemic complexity represents the amount of cognitive effort exerted by a student to pursue theoretical understanding, scientific sophistication represents the success of a student to grasp a complex scientific idea (see Zhang et al. 2009 for details). In order to assess idea improvement at the group level, four notes were identified as pivotal points where the student discussion surrounding the topic of "evaporation" became increasingly scientific. Initially, students theorized that an intrinsic property in water, such as its weight, would help it float, while external forces, such as "gravity", would pull it down. When students started discussing weather systems, such as "rain", "snow", and "hail", they started theorizing that the "sun" and "heat" played a role in evaporation. Toward the end, students had a better understanding of states of matter and started explaining the water cycle in terms of "liquids", "solids", and "gases"; some even started investigating particle theory and talked about "molecules", "hydrogen", and "oxygen". The progression of their ideas parallels curriculum units in grades 2 (air, water, environment), 5 (properties of and changes in matter), and 9 (atoms, elements, compounds), respectively.

2.3.4 Social Network Analyses

Student s1251855 was the most influential leader. Figure 2.3 shows the network analyses in KBDeX when student s1251855 was leading. The student network in Fig. 2.3a shows that student s1251855 connected students s1251934, s1251914, and s1251918 to the larger group network. The note network in Fig. 2.3b shows that note 6 written by student s1251855, linked notes 1, 9, and 10 to the larger cluster of notes. The word network in Fig. 2.3c shows that student s1251855 connected the concepts of "air", "rain", and "plant" to the main discussion of evaporation.

Below is an excerpt of the student discourse in the Evaporation view when student s1251860 is how water evaporates. Student s1251855 was the first to introduce the idea of "air" and "rain" in the discussion of evaporation, which prompted further discussion about the water cycle and weather systems.

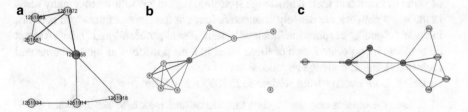

Fig. 2.3 (**a**) KBDeX visualization of student network, (**b**) KBDeX visualization of note network, (**c**) KBDeX visualization of word network at turn 9, when student s1251855 had the highest betweenness centrality

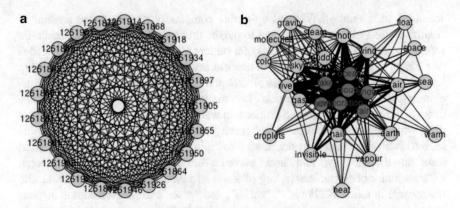

Fig. 2.4 (a) KBDeX visualization of student network, (b) KBDeX visualization of word network

s1251860:   [I need to understand]: where does water go

s1251889:   [My Theory]: I think when it evaporates it turns into steam

s1251881:   [Important Information + source]: my dad told me that evaporation can do all sorts of things. I think that everyone is right because it can do all sorts of things.

s1251855:   [My Theory]: I think that the water goes up into the "air". It "evaporates" because it needs to turn into "rain" because "plants" need water and "rain".

s1251918:   [My Theory]: well…I think that the water goes up into the sky and comes down as rain and also comes down as snow.

2.3.5   Student Portfolios

Preliminary analyses of student portfolios were conducted in KBDeX in order to examine the extent to which student ideas were connected at the end. Figure 2.4a shows that all students in the student network are connected to one another, suggesting that students acquired shared vocabulary and reached a common understanding of water at the end of their Knowledge Building. Figure 2.4b shows that many ideas in the word network are densely connected. Ideas at the center of the word network include: "cloud", "evaporation", "snow", "rain", "sun", "ocean", and "lake". Almost all students incorporated each of these ideas in their portfolios as they documented their individual knowledge advances.

Below is an excerpt from student s1251855's portfolio:

"I think the water "evaporates" up into the "clouds" and turns into "air" and when the "clouds" get too heavy with water it comes down as "rain", "snow" and "hail" and it goes into the ground and then it makes groundwater. When it's in the ground, they make "lakes" and "ponds" and "puddles" and most of the water is covering the "earth", but eventually it goes into the "ocean". And then it starts all over again."

2.4   Conclusions and Future Directions

Preliminary results confirm that this grade 1 class was indeed a successful Knowledge Building community, with students rotating leadership, actively improving ideas, all the while becoming a highly discursively connected community. Although only one example of leadership is reported here, similar results are found for the other top two leaders, and additional analyses with less influential students are underway in order to compare their learning trajectories.

The current study represents the first attempt to connect individual learning outcomes with rotating leadership at the group level during Knowledge Building. Recall that rotating leadership is an indicator of group creativity in COINs (Gloor 2005). Within the Knowledge Building context, it is interesting to note that when students assume leadership, they are connecting together unique ideas in the community knowledge, thus embodying the principles of *idea diversity*, *idea improvement*, and *rise above*. In other words, rotating leadership may be a useful way for teachers to identify students who are working creatively with ideas during Knowledge Building. It is possible that these students are facilitating the spread of improved ideas in the network or better yet, opening up new areas of inquiry for knowledge advancement. Further work is needed in order to understand how rotating leadership contributes to Knowledge Building across various contexts. Additionally, the role of the teacher needs to be examined to better understand how to facilitate self-organization around idea improvement. As in COINs, teachers and students in Knowledge Building communities are constantly redefining their membership as means to advance their community knowledge (Gloor and Cooper 2007). It is suspected that the role of teacher is to nurture swarm creativity and collaborative innovation in the classroom—a type of "disciplined" improvisation (Sawyer 2004).

In today's knowledge society, developing cultural capacity to innovate represents an educational priority. Knowledge Building pedagogy—informed by assessment designs from Collaborative Innovation Network theory—has the potential to transform classrooms into knowledge-creating communities, preparing students of all ages with competencies to communicate, collaborate, and innovate.

References

Gloor PA (2005) Swarm creativity: Competitive advantage through collaborative innovation networks. Oxford University Press, New York

Gloor PA, Cooper S (2007) Coolhunting: Chasing down the next big thing. AMACOM, New York

Kidane YH, Gloor PA (2007) Correlating temporal communication patterns of the eclipse open source community with performance and creativity. Comput Math Organ Theory 13(1):17–27

Ma L, Matsuzawa Y, Chen B, Scardamalia M (2016) Community knowledge, collective responsibility: The emergence of rotating leadership in three knowledge building communities. In: Proceedings of the 12th International Conference of the Learning Sciences, International Society of the Learning Sciences, Singapore

OECD (2015) The innovation imperative: Contributing to productivity, growth and well-being. OECD, Paris

Ontario Ministry of Education (2007) The Ontario curriculum, Grades 1–8: Science and technology. http://www.edu.gov.on.ca/eng/curriculum/elementary/scientec18currb.pdf

Oshima J, Oshima R, Matsuzawa Y (2012) Knowledge building discourse explorer: A social network analysis application for knowledge building discourse. Educ Technol Res Dev 60(5):903–921

Sawyer RK (2004) Creative teaching: Collaborative discussion as disciplined improvisation. Educ Res 33(2):12–20

Sawyer K (2015) Organizational innovation and improvisational processes. In: Garud R, Simpson B, Langley A, Tsoukas H (eds) The emergence of novelty in organizations. Oxford University Press, Oxford, pp 180–215

Scardamalia M (2002) Collective cognitive responsibility for the advancement of knowledge. In: Smith B, Bereiter C (eds) Liberal education in a knowledge society. Publishers Group West, Berkeley, pp 67–98

Scardamalia M (2004) CSILE/knowledge forum®. In: Education and technology: An encyclopedia. ABC-CLIO, Santa Barbara, pp 183–192

Scardamalia M, Bereiter C (2014) Knowledge building and knowledge creation: Theory, pedagogy, and technology. In: Sawyer K (ed) The Cambridge handbook of the learning sciences. Cambridge University Press, New York, pp 397–417

Tan SC, Tan YH (2014) Perspectives of knowledge creation and implications for education. In: Tan SC, So HJ, Yeo J (eds) Knowledge creation in education. Springer, Singapore, pp 11–34

United Nations (2005) Understanding knowledge societies in twenty questions and answers with the index of knowledge societies. Department of Economic and Social Affairs, United Nations, New York, https://publicadministration.un.org/publications/content/PDFs/E-Library%20Archives/2005%20Understanding%20Knowledge%20Societies.pdf

Zhang J, Scardamalia M, Reeve R, Messina R (2009) Designs for collective cognitive responsibility in knowledge-building communities. Journal of the Learning Sciences 18(1):7–44

Chapter 3
Creating Community Language for a Collaborative Innovation Community

Iroha Ogo, Satomi Oi, Jei-Hee Hong, and Takashi Iba

3.1 Introduction

Every community has some form of shared mindsets and visions that were formed by the founders and the people who contributed to its growth. Moreover, the behavior, attitude, and style of the community are also created by its members. These are important factors to differentiate one community from another, and also for the community to continue to exist. In the beginning stages, a community's identity is repeatedly told over and over again mainly by the founding members to be passed down. However, eventually when these veteran members have to leave the group, their unique attributes and qualities that originally existed are also lost. Moreover, it is not the mere copy and paste of the same traditions that is important, but they have to adapt to the ever-changing world it is a part of, while preserving their unique identity.

We can see examples of these communities in universities with visions and missions which were set at the time of the school's establishment. Students and teachers (whom are members of the community) establish unique mindsets and viewpoints of their university while spending their time studying their interests at the university, while contributing to shape the community's identity as a whole. The community has some of mobility because of the school changes constantly as students come in and out. In this case, the quality and identities of the school will be easily lost if such qualities are not continually updated or reproduced.

There are also other more open and self-directing communities similar to that of a Collaborative Innovation Network (COIN), where people come in and out freely

I. Ogo • S. Oi • T. Iba (✉)
Faculty of Policy Management, Keio University, Tokyo, Japan
e-mail: s14215io@sfc.keio.ac.jp; iba@sfc.keio.ac.jp

J.-H. Hong
Faculty of Environment and Information Studies, Keio University, Tokyo, Japan

© Springer International Publishing Switzerland 2016
M.P. Zylka et al. (eds.), *Designing Networks for Innovation and Improvisation*,
Springer Proceedings in Complexity, DOI 10.1007/978-3-319-42697-6_3

21

(Gloor 2010). Even in such community, there are values and ways of thinking that are handed down from person to person, which makes a community unique and attractive for others to join. In this way, it is indeed highly crucial to share the mindsets and identities of open communities in a way that can be shared.

This chapter focuses on the first type of a case—the challenge to scribe out the spirit shared and passed down at Keio University Shonan Fujisawa Campus (SFC) into words.

3.2   The Case of SFC Culture Language at SFC

At Keio University Shonan Fujisawa Campus (SFC), there are three faculties: Faculty of Policy Management, Environment and Information Studies, and Nursing and Medical Care. This campus was built to provide education to help them survive from complicated society by cooperating with people from various fields. It is the place for students to grow their personality, enhance creativity, and strengthen their way of thinking. As SFC is on the cutting edge of the environment for education and has flexible curriculum, it is used to lead the reformation of Japanese universities.

Likewise, SFC is such a distinctive campus. However, 25 years have passed from its establishment and there are only few people who actually know and understand the initial concept of this campus. Even though some of the people might know a brief history of SFC and understand its concept, it is not always possible for them to adjust it to the present circumstance with modern technology. Therefore, we created the "SFC Culture Language" which is made to reproduce the spirit of our community, SFC, in our own way.

The creating process of SFC Culture Language is divided into five steps: Interview/Case Study, Visual Clustering, Seed Writing, Culture Word Writing, and Illustration. This section explains about the writing process.

Firstly, we interviewed around 50 students, graduates, and faculty members of SFC. It took about 60 min for each interview. We also read books that were related to SFC and collected all the episodes occurred at SFC since it was established. During the interview, "What do you think characteristics of SFC is," "In what way does characteristics exist within SFC," "What do you think the advantage of characteristics is." When doing this, we wrote down all the useful information on post-its. Furthermore, find case studies from newspapers or books, and write down the information on post-its as well. The episodes repeatedly appeared in the interview or case studies are the information needed for SFC Culture Language.

After that, we clustered all the post-its we wrote by using *visual clustering method* (Iba and Isaku 2012). When doing this, we put the post-its with similar information together and group them.

The next step is seed writing. The purpose of this step is to realize what "Frequently observed phenomenon," "Background of the phenomenon," and "Significance of the phenomenon" are. These elements were written based on the main message of the groups, which were divided in the previous step, and we named

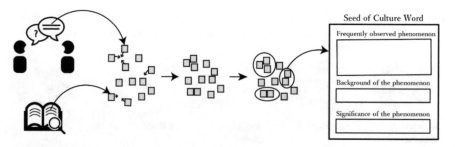

Fig. 3.1 Creating process of SFC Culture Language

to these groups "Seed of a Culture Word" (Fig. 3.1). By reading "Frequently observed phenomenon" and "Significance of the phenomenon," it encourages people to think about the phenomenon with the significance of it at the same time. Moreover, by looking at "Background of the phenomenon," people can understand how the phenomenon was created.

Once seed writing is completed, we added or edited contents in order to make them easy to read. We called this finished "Culture Word". During this process, we revised Culture Word over and over in order to raise the quality of it. Then, to use these Culture Words in conversation, we named it. Lastly, we draw an illustration based on the main and the most important message of the Culture Word.

This SFC Culture Language was handed out about 900 including both Japanese and English version at SFC 25th Anniversary Ceremony. Figure 3.2 shows all 27 Culture Words of SFC Culture Language, and three examples are shown in Figs. 3.3, 3.4, and 3.5.

3.3 Using SFC Culture Language in Dialogue Workshop

The words in the SFC Culture Language were created by applying the method of Pattern Language (Alexander 1979) to make them usable in conversations. In this section, we will show the results from a dialogue workshop (Iba 2012) we held using the language which was held with 25 SFC students on 14 January 2016.

In this workshop, we asked two main questions. At first, we asked the participants to read SFC Culture Language in advance, and check all the characteristics they have experienced before. Furthermore, we asked participant to leave a comments on it. Through this work, we could check whether the identity exists in the community; SFC, is well described in the culture language or not. The results are below. Comments on SFC Culture Language are "We could agree with each other," "It was possible to think carefully about SFC once again," "There were many SFC-like cultures in SFC Culture Words and I wanted to show them to graduates," "I think these Culture Words show SFC's identities." According to the checklist, about 70 % of the participants answered that 24 out of 27 Culture Words were seen at SFC. From both comments and checklist, we could assume that SFC Culture Language contains cultures actually exist in SFC.

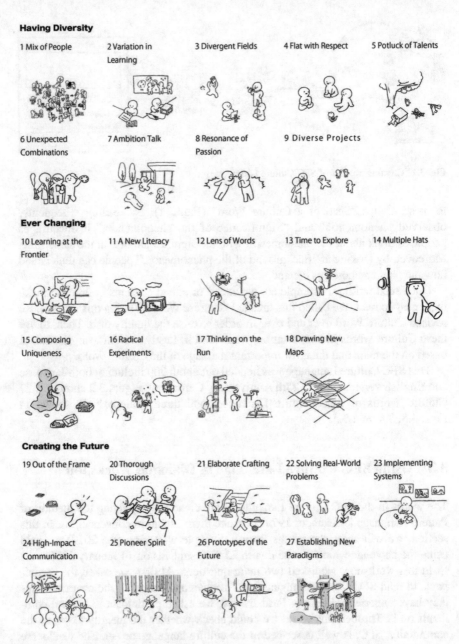

Fig. 3.2 Culture words and its illustration of SFC Culture Language

In addition, we asked participants to find a person who has already experienced the characteristics they want to experience. They could listen to that person's story, and think how to put that culture into their life. As people used Culture Words in their

Unexpected Combinations

Form a group with an unusual combination of fields. Work in a team of people from completely different fields. For example, combine people who are studying different research areas: the Internet and economical policy, architecture and sociology, work in a team of experts from Biology, Neurological Science, Pharmacy, Communication and Policy under the topic of "Health". Such combinations of people are likely to occur at SFC with people of different study fields sharing the same campus.

▲ System and / or the way of thinking that causes this

Many of the faculties have very unique research topics, exchange of knowledge between the different areas occur on a daily basis. For students, with a Variation in Learning among them, they can meet various types of people through group work in classes. With such a diverse environment, new combinations are very likely to occur. With a Flat with Respect relationship and a Potluck of Talents, this synergy is even more enforced.

▼ Therefore

A new research or project might emerge. By having people with diverse knowledge interacting with each other, we become able to understand and analyze problems from different perspectives. This would help create a synthetic solution that would not have been found if done alone. Moreover, with several fields merging together, a completely new idea may be generated. The "Socio-semantics" is an example of this with SFC professors with a political science and a linguistics background combined their studies.

Fig. 3.3 An example of SFC Culture Language: unexpected combinations

conversation during this workshop, we could realize that this language helps them to understand the identity even better, and reconsider SFC Culture Language on one's own context. The answers we got from the survey are below. "As there are words for each culture, we could realize the existence of the culture and made us want to try it," "I could realize that there is some culture I have never experienced before even though

A New Literacy

Discover skills or technology that would become essential for the coming future. "Reading, writing and arithmetic" - these were for a long time considered as the essential skills a person must have. However, times have changed and now these aren't enough to create new values. At SFC, we define "feeling, thinking and learning" as the set of new literacy that would start to become important. At the time when the campus was first founded, "artificial language, natural language and the three techniques (social research law, model simulation techniques and multivariate analysis)" were considered as the three main study areas. Since then we have defined various skill sets that are important for the specific point of time. These include design language, knowledge skills, and data science. More recently, digital fabrication skills are starting to become important, and an environment to acquire these skills are set up at SFC. The literacy changes according to the time but these skills have always been reflected onto the curriculum throughout time.

▲ System and / or the way of thinking that causes this

In addition to obtaining the newest knowledge, the education and research of creation and practice skills are also unique characteristics of SFC. When deciding on a new curriculum, several faculties gather to think what kind of an education system would be the most "SFC-ish." They suggest essential subjects that students should learn at the committee meetings. A similar process is taken for determining on new facilities on campus. While the students learn new skills or knowledge in class, they can also teach other students as an SA (Student Assistant) or as a consultant at the Media Center to brush up their skills even more.

▼ Therefore

By using the new technology or skills, you can broaden the idea of what you can do. Students will not only "know" about the technology, but will also learn how to "utilize" it as well. This way you can use the technology to extend your own ideas. For instance, when you face a problem, if you have some fabrication skills, you have one more option for a solution to the problem. With the new literacy, you will have a sense of thinking what kinds of solutions there are, and which ones are realistically feasible. By adapting the literacy that suits the trend, you would be able to build up skills that would become useful anywhere in the world.

Fig. 3.4 An example of SFC Culture Language: a new literacy

Establishing New Paradigms

Create new methods and systems of study with the establishment a new paradigm of study in perspective. When we are Solving Real-World Problems, and Implementing Systems to conduct research, this is not just any social practice. We are trying to create new methods that could possibly establish a new paradigm of study. Some of the leading disciplines that people study today were not considered a discipline just a few centuries ago. For example, it was only in the 19th century that, economics, sociology, and politics split into different disciplines. The establishment of the new studies into a science is something that can happen, and at SFC this possibility is always thought about.

▲ System and / or the way of thinking that causes this

SFC is not a place where students are taught an existing paradigm. Faculties do not stick around to the existent field but are recommended to create new fields. In reality, if you are researching something Out of the Frame, you must be able to define and explain what you are researching, but somewhere in your mind you should have the ambition to nurture it into a science of its own. Furthermore, it is not unusual to see faculties researching seemingly unrelated topics coming across a common problem, and collaborating to find a new area of study. Also, students are able to join more than one research seminars. Therefore, students can find their own new boundaries that the faculties could not thought up with by combining multiple areas of study.

▼ Therefore

In SFC, students study in new research and study fields that are Prototypes of the Future. Moreover, we build innovative developments that cannot possibly be done without the mind of Establishing New Paradigms. They were made possible intellectual environment particular at SFC, including the developing technology, evolving of methods, understanding histories, accepting diverse cultures, and practicing policy management, etc. While Solving Real-World Problems is progressing, at the same time, you are Establishing New Paradigms with a Pioneer Spirit and contributing to create the future.

Fig. 3.5 An example of SFC Culture Language: establishing new paradigms

I am still studying at SFC," "It was interesting to talk and think about the system needed to experience the culture of SFC," "If the culture that people really want or need is well written in the Culture Words, it would be useful."

The results of the workshop show that SFC Culture Words could be used for understanding or for reconsidering the characteristics. However, the workshop we held was an experimental workshop, and the number of participants is too small to make a concrete conclusion. We would like to hold this workshop again with more people to get a valid result.

3.4 Conclusion

In this chapter, we introduced our challenge to create a method for communities reproduce its own identity while taking in new ideas. The challenge involved scribing out the spirits, visions, habits, attitudes, and senses shared among the community in small unit words usable as common language. We thought that this would enhance the chain of communications within the community. As our first case, we took an example for university faculty and campus though; we think this method also can be used for open community. Our future prospect is to make more cases in other fields.

Acknowledgments Our great thanks go to all of our interviewees, workshop participants, and the members who helped us publishing our booklet, "SFC Culture Language: Words that express qualities of SFC." We would also like to thank our project team member, Yu Tiffany Morimoto, and the members of Iba laboratory, especially Yuji Harashima, Norihiko Kimura, and Tetsurou Kubota, who gave us great comments and advices. We also appreciate the support of Taichi Isaku and Ayaka Yoshikawa who gave us much advice and translated our paper into English.

References

Alexander C (1979) The timeless way of building. Oxford University Press, New York
Gloor P (2010) Coolfarming: Turn your great idea into the next big thing. AMACOM, New York
Iba T (2012) Dialogue workshop patterns: A pattern language for designing workshop to introduce a pattern language. Paper presented at the 17th European Conference on Pattern Languages of Programs, Kloster Irsee, Germany, 11–15 July 2012
Iba T, Isaku T (2012) Holistic pattern-mining patterns: A pattern language for pattern mining on a holistic approach. Paper presented at the 19th Conference on Pattern Languages of Programs, Tucson, Arizona, USA, 19–21 Oct 2012

Chapter 4
Sociological Perspective of the Creative Society

Takashi Iba

4.1 Introduction

To imagine what a society of the future might be like, we often look at the transitions between the past, present, and future. We look at the flow from the past to the present, and then imagine how the flow may extend into the future. One way to look at past and present transitions of society is to view the transition from the consumptive society to communicative society. Earlier there used to be only end consumers of products; however, today consumers send out information about the product from their own experiences. From this transition, one could extrapolate a next age to be full of people each creating goods for their own consumption that could be called an emerging stage of creation (see Fig. 4.1).

If a subsequent transition is observed from a time of mass consumption to a time of greater personal consumption and related communication, there could be a transition from just plain communication of ready-to-go information to a time of creation. This new kind of society is part of my vision of a Creative Society (Iba 2013). People of a Creative Society could create their own goods, tools, concepts, knowledge, mechanisms, and ultimately, the future with their own hands. Creation in this society would no longer be limited to just companies and organizations but could be performed by each and every individual according to their own satisfaction.

This chapter presents theoretical consideration of the Creative Society with using the social systems theory and creative systems theory.

T. Iba (✉)
Faculty of Policy Management, Keio University, Tokyo, Japan
e-mail: iba@sfc.keio.ac.jp

© Springer International Publishing Switzerland 2016
M.P. Zylka et al. (eds.), *Designing Networks for Innovation and Improvisation*,
Springer Proceedings in Complexity, DOI 10.1007/978-3-319-42697-6_4

Fig. 4.1 An overview of
social change

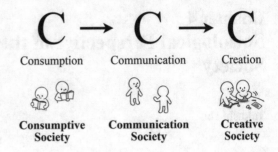

4.2 Systemic Viewpoint of Society

Niklas Luhmann (1927–1998) was a sociologist who persistently studied the various domains of society and built what is called as a grand theory of society. He tackled fundamental questions regarding how societies are possible. His most innovative contribution was the application of the most current systems theory to sociology and adopting the theory of autopoietic systems to understand modern societies. Autopoietic systems were proposed by Humberto Maturana and Francisco Varela in biology (Maturana and Varela 1972) as a unity where the organization is defined by a particular network of production processes of elements. Although the original area of autopoietic systems is biology, Luhmann generalized the theory and used it to enhance his theories of social systems as autopoietic systems that include communication as an element.

The social systems theory considers society as an autopoietic system whose element is communication. Social system is reproduction networks of communication. Communication can have no duration because it is momentary operation, so it must be reproduced constantly. From an operations viewpoint, such a social system is a closed system as it cannot send or receive communication outside the system.

The theory also considered the human mind as an autopoietic system, which is called as a "psychic system," where the primary element is consciousness. Psychic system is a nexus of consciousness, and the system reproduces consciousness by consciousness. The consciousness can have no duration because of its momentary operation and requirement to be reproduced constantly. Moreover, from an operational viewpoint, such a psychic system is a closed system as it cannot send or receive communication outside the system. Psychic systems are mutually inaccessible, thus making some type of communication necessary.

To formulate interactions between autopoietic systems is much more complex than with conventional systems theory because the autopoietic system is operationally closed. An autopoietic system cannot recognize other systems because there is a distinction between its own system and its environment. The concept of "structural coupling" is introduced in contrast to operative coupling to describe the relationship between function systems in autopoietic systems theory.

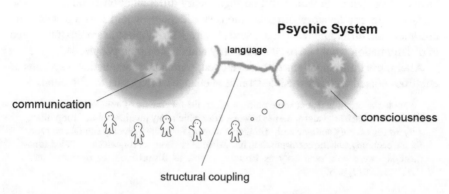

Fig. 4.2 Social system and psychic system in Luhmann's Social Systems Theory

Languages as media provide the chance of coupling between consciousness and communication using symbolic generalizations for mutual comprehension. Language is a means of communication and also thinking, as Luhmann declared, "linguistically formed thoughts play a part in the autopoiesis of consciousness, help to produce it" (Luhmann 1995). This is an overview of the most fundamental framework proposed by Niklas Luhmann. The concepts and their relations are shown in Fig. 4.2, and they are basic viewpoints of the world that I adopt in this chapter.

4.3 A Theory of Modern Society

Luhmann advanced his thoughts on modern society as an autopoietic system consisting of several functional systems. Some examples are as follows: economy, law, politics, art, science, and religion. Understanding Luhmann's point of view is crucial to understand the Creative Society of this chapter.

A reduction in overwhelming complexity is necessary for society to maintain its social order. Luhmann proposed some forms of differentiation in the history of society: segmentary differentiation, center–periphery differentiation, stratificatory differentiation, and functional differentiation. Luhmann considered modern society as a functional differentiated society (Luhmann 1989, 2013). He posited that in the societies' primitive stages, it was "divided into basically similar subsystems, which mutually constitute environments for one another" (Luhmann 2013, p. 27). At its simplest, there were two levels: families and society (horde). However, at three levels, they would be: families, villages, and tribes. In a society with segmentary differentiation, territory and kinship relations defined the boundary of its subsystems, and the social position of individuals was fixed over time.

If some social catastrophe occurred, it would trigger a rapid transition in such a system, and center–periphery differentiation and stratificatory differentiation would develop. In center–periphery differentiation, the centers would differentiate and have closure, but there would still be segmentary differentiation of families in the periphery. In stratificatory differentiation, stratification would develop when the upper stratum differentiated and reached closure. The simplest example of this type of differentiation would be between nobility and the common people.

After these differentiations fully developed, Luhmann saw functional systems as outdifferentiated with irreversible structural changes. According to Luhmann,

> "What is important is that at some point or other, the recursivity of autopoietic reproduction began to take hold and achieved closure, after which only politics counted for politics, only art for art, only aptitude and willingness to learn for education, only capital and profit for the economy, and the corresponding intrasocietal environments—which included stratification—were now seen only as irritating noise, as disturbances or opportunities." (Luhmann 2013, p. 66)

4.4 Functional Systems

Luhmann (1989, 2013) listed functional systems of modern society as follows: economy, law, science, politics, art, education, religion, mass media, medical care, and family (Fig. 4.3). Table 4.1 provides a summary of the characteristics of these

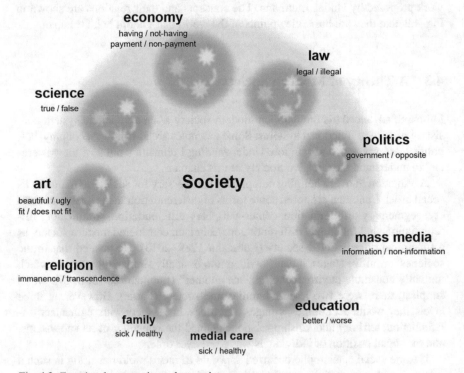

Fig. 4.3 Functional systems in modern society

Table 4.1 Features of functional systems in modern society, discussed by Niklas Luhmann

System	Function	Code	Program	Symbolically generalized media
Economy	Ensuring future supply under the condition of scarcity (Luhmann 2013)	Having/not-having Payment/non-payment	Prices, investment programs, liquidity assessment programs	Money
Law	Stabilizing normative expectations (Luhmann 2008)	Legal/illegal	Legal norms (laws, previous legal decisions, etc.), courtroom proceedings	
Science	Acquiring and generating new knowledge (Luhmann 1989)	True/false	Theories, methods	Truth
Politics	Enabling to enforce binding decisions collectively (Luhmann 1990)	Government/opposition		
Art	Providing new ways of observing society by pointing to an imaginary order (Luhmann 2000a)	Beautiful/ugly Fit/does not fit	Manifests, styles	Art which can be expressed in the form of paintings, sculptures, installations, etc.
Education	Changing people to prepare for communication that takes place in society (Luhmann 1989)	Better/worse, in performance	Curricula and readings	
Religion	Translating the indeterminacy of the world into a meaningful order (Luhmann 1989)	Immanence/transcendence		Faith
Mass media	Contributing to construction of reality in the society (Luhmann 2000b)	Information/ non-information	News and in-depth reporting, advertising, entertainment	
Medical care		Sick/healthy		
Family		Love/no love		Love

functional systems. First, each function of the social subsystems works for a specific societal problem and they are different from one another (Table 4.1, second column). For example, economy works to regulate scarcity and ensure future supply, a science system fulfills the function of acquiring and generating new knowledge, and politics realizes the enforcement of order on people as decisions are made collectively. Every functional system has its own code and program to realize its autopoietic system operations and has some type of media for supporting communication.

4.4.1 Code and Program

Each functional system carries out its operations of autopoiesis through a binary code, two-sided form as a way to describe society, that can be seen as simply code (Table 4.1, third column) such as: having/not-having properties; payment/non-payment in an economy; legal/illegal in the law; true/false in science; government/opposition of authority in politics; originally beautiful/ugly and then fit/does not fit in art; better/worse performance in education; immanence/transcendence in religion; information/non-information in mass media; sick/healthy in medical systems; and love/no love in families.

Functional systems structure communication within according to their own codes. All communication relates to a code belonging to the system of that code. For instance Luhmann says, "The legal system receives its autopoiesis through coding the difference between what is legal and what is illegal" (Luhmann 1989, p. 64) and "The difference between true and false is what matters for science's code" (Luhmann 1989, p. 76). Abstractly speaking, the codes always take the form of A/not-A. Code is the key mechanism of differentiation in modern society according to Luhmann, "For a description of modern society one will have to admit that important and distinctively modern function systems have become identified through a binary code that is specifically valid for each of them" (Luhmann 1989, p. 42).

This concept of binary code may look too simple to describe the fundamental mechanisms of a modern society, but actually, it can work well universally and sustainably because of its simplicity. This is one of the key points of great interest in Luhmann's social systems theory. An important concept in the social systems theory is programs to express the decision rules to assign each target to a positive side or a negative side of the binary code (Table 4.1, fourth column). For example, in the legal system, the distinctions between what is legal and what is illegal are carried out based on legal norms such as existing laws and previous legal decisions. These legal norms are programs of the legal system. In science, theories and methods are used as programs to determine the difference between true or false. Another example that is easy to understand are the programs of mass media, which structure their communication code of information/non-information based on several programs such as news and in-depth reporting, advertising, and entertainment. In the social systems theory, code is strictly fixed, whereas, programs can change over the long term. This combination of unchangeable and changeable is a secret in managing the complexity and openness of further possibilities.

4.4.2   Communication Media

In social systems, there are always three uncertainties: uncertainty of understanding others, uncertainty of achievement, and the uncertainty of the results of communication, due to the difficulty in generally understanding what others are thinking since psychic systems are operationally closed to others (Luhmann 1995). Since there is uncertainty for realizing communication, it is intrinsically difficult to realize the nexus of communication.

However, in reality, some kind of evolutionary achievement, called media, provides support to overcome the uncertainty of understanding others. Languages, as such media, provide the chance of coupling between consciousness and communication using symbolic generalizations to produce mutual comprehension. Language is not only a means of communication but also critical to thinking.

To act against the uncertainty of achievement dissemination media exists. Typical examples of dissemination media are newspapers, television, and the Internet. Communication and media studies generally focus on language and dissemination media.

Furthermore, there is a third media that works against the uncertainty of the results of communication, which is known as symbolically generalized communication media. Prime examples are love, power, and money (Table 4.1, fifth column). These media activate the motivation of people to participate in communication such that there are successful results in accepting the meaning of communication. Although functional differentiation is a key mechanism in modern society, symbolically generalized communication media is another important key that makes modern society possible. In this regard, Luhmann states, "it should be noted that symbolically generalized communication media are suitable only for functional areas in which the problem and the success aspired to lie in communication itself. The function is fulfilled if the selection of a communication is taken as the premise for further communications." (Luhmann 2012, p. 246).

4.4.3   Structural Coupling

According to social systems theory, functional systems relate to each other by way of structural coupling. Structural coupling is a way of creating influence between systems without having common elements. This concept is quite important because autopoietic systems are by definition, operationally closed, thus unable to interact or share something. Therefore, functional systems may influence each other via structure, which regulates the scope of possibility in communication.

Examples of structural coupling as depicted by Luhmann are shown in Fig. 4.4. Structural coupling is achieved through the following: taxes and charges between politics and the economy, constitutional authority between politics and law, property and contract between economy and law, references and certificates between economy and education, universities between science and education, expert advice between science and politics, advertising between economy and mass media,

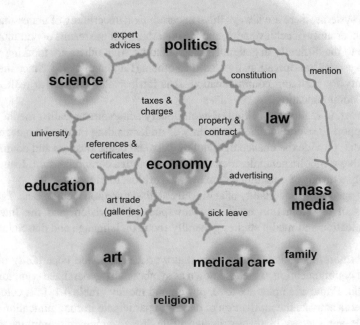

Fig. 4.4 Structural couplings between different functional systems

references between politics and mass media, art trade (galleries) between art and economy, and sick leave between medical care and economy. Thus, each functional system works for its own specific function to realize the society while loosely influencing one another, but without central control.

4.5   Co-Creation as a Functional System

Based on the theoretical framework, I now redraw a map of modern society as an age of creative society. In my view, the creative society is in modern society, not post-modern, because even in creative society, functional differentiation is still at work as economy, law, and science. Therefore, characteristics must be recognized within the framework of social systems theory that supports a creative society. My proposal is that a co-creation system as a functional system of the current society is emerging (the creative society). As Luhmann studied the emergence of functional systems in history, we can imagine a vision of new functional systems of the co-creation system as supplying creativity, especially in collaborative ways (Fig. 4.5, Table 4.2, second column).

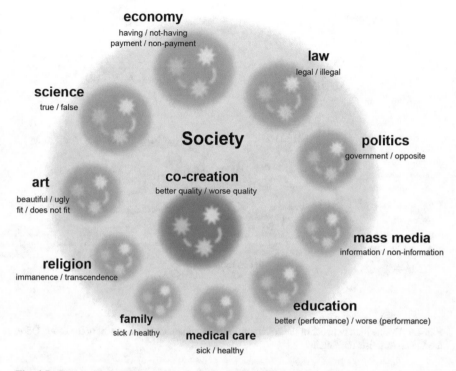

Fig. 4.5 Co-creation system as a new functional system in a creative society

Table 4.2 Features of the functional system proposed in this chapter

System	Function	Code	Program	Symbolically generalized media
Co-creation	Supplying creativity, especially in a collaborative way	Better/worse, in quality	Patterns in pattern languages	Pattern languages

Figures 4.6 and 4.7 show co-creation network in the case of open-source development of Linux OS in Linux-Activists mailing list and comp.os.minix Newsgroup respectively (Iba 2007). These networks are visualized with investigating the post and responses. Also, Fig. 4.8 shows co-creation networks as people edit online articles in Wikipedia (Iba et al. 2011). These networks are visualized based on the data of sequential order to edit the target articles. These cases are examples of co-creation on the Internet. However, in general, such a continuous reproduction of communication is difficult to maintain with a few exceptions. Therefore, further examination of how co-creation systems are emerging is needed.

Luhmann asserted, "Social systems arise through the initiation of communication, and develop autopoietically from within themselves" (Luhmann 2013, p. 341). Accordingly, he claimed, "My position is that binary codes having these characteristics occur in social evolution and that, if they are put into operation, corresponding

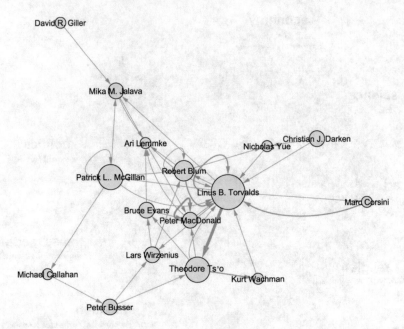

Fig. 4.6 Co-creation network of the collaboration of open-source development of the Linux OS in the Linux-Activists mailing list

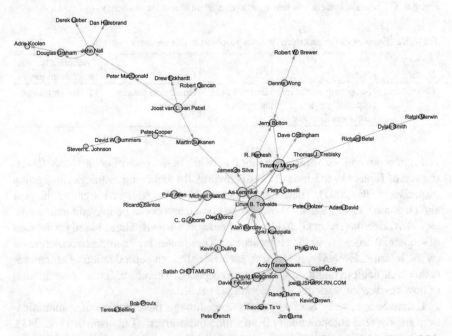

Fig. 4.7 Co-creation network of the collaboration of open-source development of the Linux OS in the comp.os.minix Newsgroup

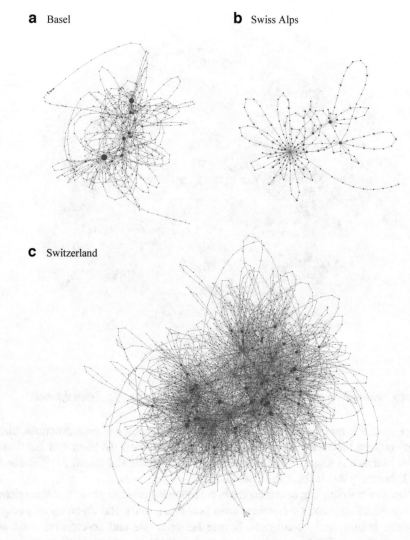

Fig. 4.8 Co-creation network of the collaboration of people editing articles in Wikipedia (Iba et al. 2011). (**a**) Basel, (**b**) Swiss Alps, (**c**) Switzerland

systems tend to be differentiated" (Luhmann 1989, p. 43); therefore, there is a need to identify the code for a co-creation system.

I propose better/worse for quality (Table 4.2, second column) as the candidate for the program of the co-creation system based on patterns in pattern languages (Table 4.2: forth column). Pattern languages are collections of patterns that describe practical knowledge in a certain domain and define *what is good*. The original idea of pattern languages was proposed in architecture (Alexander et al. 1977) and later applied to software design (Beck and Cunningham 1987; Gamma et al. 1995).

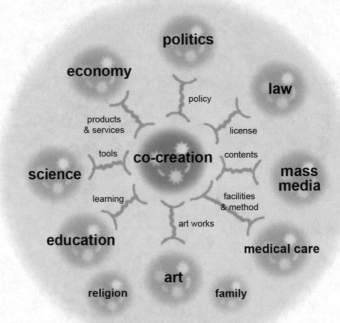

Fig. 4.9 Structural couplings between a co-creation system and other functional systems

Since then, this method has been applied to describe a range of human actions, and many pattern languages have been created and published (Coplien and Harrison 2004; Manns and Rising 2004; Pedagogical Patterns Editorial Board 2012; Iba and Iba Laboratory 2014a, b, c; Iba et al. 2015).

Structure coupling can be considered in the following manner: products and services between economy and co-creation, tools between science and co-creation, policy between politics and co-creation, license between law and co-creation, content between mass media and co-creation, art work between art and co-creation, facilities and method between medical care and co-creation, and learning between education and co-creation (Fig. 4.9).

4.6 Structural Coupling Between Social Systems and Creative Systems

While discussing the emerging functional system of a creative society from the social side, another key system to make a creative society possible is a creative system, which is not a social system but another type of autopoietic system. I have proposed a creative system to describe creative processes (Iba 2010). In this creative

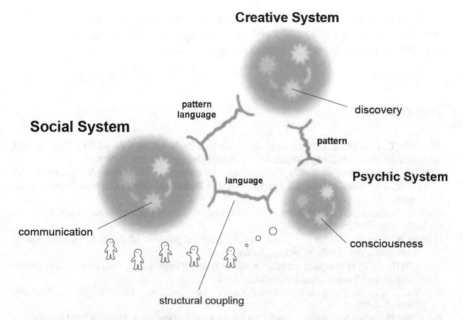

Fig. 4.10 Structural couplings among social systems, psychic systems, and creative systems

systems theory, a creative process is defined as an autopoietic system where elements are discoveries. In other words, a creative process is a reproduction network of discoveries. This theory also defines each discovery as emerging only when a synthesis occurs between idea, association, and finding.

I want to emphasize that there is a difference between a co-creation system as a social system and a creative system that describes a creative process without social aspects. Note that a creative process can be carried out by one person without any social interaction, which is why a creative system is required to be considered separately from a co-creation system on the social side. Co-creation systems and creative systems are different types of autopoietic systems, which can work collaboratively to realize socially creative activities.

In a past study (Iba 2010), the structural coupling of social and creative systems was described as achieved with pattern language (Fig. 4.10) between a creative system and psychic system that informs society of a way to solve a certain problem. In addition, as I mentioned earlier in this chapter, a social system and a psychic system are structurally coupled by languages, including pattern languages. Thus, you can see pattern languages are a key to structural coupling between systems for a creative society.

4.7 Conclusions

Alan Kay, the founding father of personal computers, says, "The best way to predict the future is to invent it." In this chapter, I used sociological theory to build a theoretical perspective of a future society. Although this is an unusual usage in sociology,

I believe that this kind of attempt is important to realize an ideal future creatively. However, the perspective proposed in this chapter is just a projection that will hopefully promote future discussions along these lines.

References

Alexander C, Ishikawa S, Silverstein M, Jacobson M, Fiksdahl-King I, Angel S (1977) A pattern language: Towns, buildings, construction. Oxford University Press, New York

Beck K, Cunningham W (1987) Using pattern languages for object-oriented programs. In: OOPSLA-87 Workshop on the Specification and Design for Object-Oriented Programming, Orlando

Coplien JO, Harrison NB (2004) Organizational patterns of agile software development. Prentice-Hall, Upper Saddle River

Gamma E, Helm R, Johnson R, Vlissides J (1995) Design patterns: Elements of reusable object-oriented software. Addison-Wesley, Boston

Iba T (2007) Social system analysis of open collaboration: Reconsidering open source development (in Japanese). J Infosocionomics Soc 2(2):34–51

Iba T (2010) An autopoietic systems theory for creativity. Procedia Soc Behav Sci 2(4):6610–6625

Iba T (2013) Pattern languages as media for the creative society. In: 4th International Conference on Collaborative Innovation Networks (COINs2013), Santiago

Iba T, Iba Laboratory (2014a) Learning patterns: A pattern language for creative learning. CreativeShift Lab, Yokohama

Iba T, Iba Laboratory (2014b) Presentation patterns: A pattern language for creative presentations. CreativeShift Lab, Yokohama

Iba T, Iba Laboratory (2014c) Collaboration patterns: A pattern language for creative collaborations. CreativeShift Lab, Yokohama

Iba T, Matsuzuka K, Muramatsu D (2011) Editorial collaboration networks of Wikipedia articles in various languages. In: International Conference on Collaborative Innovation Networks 2011 (COINs2011), Santiago

Iba T, Okada M (eds), Iba Laboratory, Dementia Friendly Japan Initiative (2015) Words for a journey: The art of being with dementia. CreativeShift Lab, Yokohama

Luhmann N (1989) Ecological communication. University of Chicago Press, Chicago

Luhmann N (1990) Political theory in the welfare state. De Gruyter, Berlin

Luhmann N (1995) Social systems. Stanford University Press, Stanford

Luhmann N (2000a) Art as a social system. Stanford University Press, Stanford

Luhmann N (2000b) The reality of the mass media. Stanford University Press, Stanford

Luhmann N (2008) Law as a social system. Oxford University Press, Oxford

Luhmann N (2012) Theory of society, vol 1. Stanford University Press, Stanford

Luhmann N (2013) Theory of society, vol 2. Stanford University Press, Stanford

Manns ML, Rising L (2004) Fearless change: Patterns for introducing new ideas. Addison-Wesley, Boston

Maturana H, Varela F (1972) Autopoiesis and cognition: The realization of the living. Reidel, Holland

Pedagogical Patterns Editorial Board (2012) Pedagogical patterns: Advice for educators. Joseph Bergin Software Tools, San Bernardino

Part II
Machine Learning, Prediction and Networks

Chapter 5
Predicting 2016 US Presidential Election Polls with Online and Media Variables

Veikko Isotalo, Petteri Saari, Maria Paasivaara, Anton Steineker, and Peter A. Gloor

5.1 Introduction

Preelection polls are easily available to voters (Sinclair and Plott 2012), thus assisting uninformed voters to choose their most preferable candidates without obtaining much information about either candidate (McKelvey and Ordeshook 1985, 1986). Increase of publicly available online data about political candidates makes it possible to obtain information about candidates' prospects, which could be used to predict elections through, as in this chapter, the polls themselves.

According to Zajonc's exposure theory, continuous exposure to an object often results in a positive attitude toward that object (Zajonc 1968). It has been found that younger voters' opinions are more media dependent, as they seek to set their opinions via information provided by broadcasting corporations, in comparison to older voters (Fox et al. 2007).

Presidential candidates are pushed to compete for limited media space (TV airtime and newspaper articles) in order to shape their campaigns and possible outcomes of the race (Dalton et al. 1998). The 2008 US presidential election foretold the rise of social media in political campaigning (Vitak et al. 2011). Politicians use their Facebook profiles and Twitter to mobilize support inspiring either votes or funding, which happens on social media on a large scale and with a vast audience (Dalsgaard 2008). A survey conducted in 2014 reports one-third of Facebook users post about

V. Isotalo (✉) • P. Saari • M. Paasivaara
Aalto University, Helsinki, Finland
e-mail: veikko.isotalo@aalto.fi

A. Steineker
University of Cologne, Cologne, Germany

P.A. Gloor
MIT Center for Collective Intelligence, Cambridge, MA, USA
e-mail: pgloor@mit.edu

© Springer International Publishing Switzerland 2016
M.P. Zylka et al. (eds.), *Designing Networks for Innovation and Improvisation*,
Springer Proceedings in Complexity, DOI 10.1007/978-3-319-42697-6_5

government and political issues, while 25 % of Twitter users tweet about the same issues (Pew Research 2015). However, researchers fully admit that social media users are not perfectly representing the actual general population (An and Weber 2015).

Yasseri and Bright's (2016) study on 2009 and 2014 European Parliament elections suggests that turnout can be predicted, analyzing the increase of Wikipedia page views in a period of time approaching election day. Raise in the level of Wikipedia traffic about political parties suggested successful emergence of new parties, and anti-establishment parties, while the traditional news media was a good indicator for well-established parties (Yasseri and Bright 2016).

Erikson and Wlezien (2008) claim prediction of the betting market catches up to poll projections usually by the end of the election race. Berg et al. (2001) suggest a reciprocal relation between polls and bets, as prediction market participants simply use published polls to guide them in placing bets.

5.2 Methods

This chapter seeks to answer following questions:

RQ1 Can online and media variables be used in predicting the public opinion gathered by pollsters?

RQ2 Are there statistically significant correlations between polls, Internet traffic, social media, and traditional media variables?

In this chapter, we examined nine presidential candidates in the 2016 US primary elections. The data we used comprised 13 different variables available online. The variables were chosen on the basis of their accessibility, reliability, and existence in academic literature for election predictions. Data collection was conducted in the time period from November 1st 2015 to January 14th 2016 (see Table 5.1). All data except betting was normalized.

First, we prepared the data for machine learning purposes. We assumed that there would be a lag in polls, but for instance on Twitter the public opinion is expressed in real time. We tested different combinations with polls and independent variables, and developed a formula, where days 1–3 of the independent variables were explaining the polls of the day 5. A single data point of the independent variable was thus an average of 3 days' values. Later, the results of the prediction are evaluated by examining the correlations of change variables that use the same day combination as the prediction.

Linear regression analysis along with data partitioning and scoring were done using KNIME (Berthold et al. 2007). Two models were created: a comparison model that had only lagged poll numbers and the candidate control variables as independent variables, while the prediction model would also include the various media and Internet traffic variables, along with betting as a variable. These models were compared by scoring them, using the same data set for each, with 70 % of the data in the training set and 30 % in the validation using linear sampling.

Table 5.1 Variables and their sources

Variable	Source/software	Notes
Polls	Realclearpolitics.com (2016)	Republican and Democratic national polls, linearly interpolated
Tweets	Twitter Streaming API (Dev. twitter.com 2016)	Tweets containing specific keywords, using track function
Sentiment	Stanford NLP Java library (Stanfordnlp.github.io 2016)	Average sentiment was calculated from more than 1000 tweets daily which were evenly distributed for individual candidates
Celebrity Tweets and Celebrity Sentiment		Twitter's verification parameter was used to separate celebrity tweets
Twitter Followers	Twittercounter (2016)	Daily changes in candidates' Twitter follower count
Article Mentions	The GDELT Project (Analysis. gdeltproject.org 2016)	
TV Mentions	The GDELT 2016 Campaign Television Tracker website (Television.gdeltproject.org 2016)	
Facebook Page Likes	Presidential-candidates.insidegov. com (2016)	
Talking About This	Presidential-candidates.insidegov. com (2016)	Facebook metric tracking users' activity
Wikipedia Traffic	Stats.grok.se (2016)	Page views of candidates' personal Wikipedia pages
Google Trends	Google Trends (2016)	The trend data was filtered by covering only searches inside United States and focusing on web searches. Candidate's verified search term profiles were used
Betting	Oddschecker.com (2016)	We chose to focus on individual betting firm called Betfair Exchange, which updated candidates' ratios frequently

5.3   Results

A machine learning model with linear regression was implemented and proved to be useful for predicting polls, thus providing a positive answer to our first research question. As root mean squared error adjusted for degrees of freedom gives the standard error of the regression, with 0.0058 RMSE when using 2 days old poll data in the training set, we get an error margin of roughly +/−1.2 % with a 95 % confidence interval.

The prediction model's prediction accuracy was compared to the comparison model that used only lagged poll numbers and the candidate name as independent variables. The prediction model gave roughly 10 % lower root mean squared error, along with better fitting of the data when 2 days lagged poll data was used. This advantage was there irrespective of how much the polls were lagged, but was more pronounced with more lag (Table 5.2).

Table 5.2 Prediction model and comparison model results

Lag to polls	RMSE	RMSE (Comp.)	Difference (%)	R^2	R^2 (Comp.)
Polls(-1)	0.00398	0.00408	2.41	0.999532	0.999509
Polls(-2)	0.00580	0.00631	8.78	0.999008	0.998823
Polls(-3)	0.00724	0.00844	16.4	0.998453	0.997897

RMSE and R^2 are based on prediction model including all the variables; *RMSE (Comp.)* and R^2 *(Comp.)* only use lagged polls and candidate control variable. Difference is how much larger RMSE of comparison model is to the RMSE of prediction model

To answer our second research question, we calculated the Pearson correlation coefficients between collected variables. The correlation coefficients were calculated for the changes in normalized data (except betting odds); see Table 5.3. The data set consisted of 74 instances of each variable for all candidates. These values were calculated from the normalized data by subtracting the value of the previous day from the value of the current day. Even though the correlations with polls are rather small, both betting and Facebook page likes are statistically significant at $p < 0.01$. Other variables were not statistically significant in relation to change in polls.

In the data, there is some indication of collinearity. Correlations are rather high between multiple independent variables. There is evidence of strong connections between social media and traditional media (e.g., high tweet numbers can be a result or a cause for high article mentions). Similar effects are noted for Internet traffic variables and other media. Internet traffic of the candidate is in a positive relation to other media publicity, including social media and traditional media (TV mentions and news articles).

Significant differences were found among the candidates and their polls which correlated quite differently with the changes in other collected variables. Two figures below (see Figs. 5.1 and 5.2), small fraction of data collected, present the Wikipedia traffic variable for both parties and their candidates. The figures depict differences between candidates, also marking single events on candidates' campaigns that can be clearly distinguished from bigger trends.

Ben Carson raised public attention on November 4th as it was revealed that his past had been altered for improving his media image (Daisey 2015). The peak, visible in the graph, indicated a clear negative event in Carson's campaign. In contrast as Donald Trump's Wikipedia traffic reached its peak on December 9th, his position in the polls did not encounter any negative consequences. The peak in his Wikipedia traffic was a result of taking a stand on Muslim immigration 2 days before (Johnson and Weigel 2015), proposing to shut down Internet the day before (Griffin 2015), and being attacked by a bald eagle (Guff 2015). On the Democratic side, the presidential debates sparked the most Wikipedia views on November 14th and December 19th (Democrats.org 2016).

Those previous examples about Wikipedia traffic alleviate the differences in reasons which cause variables to increase and decrease and that the changes in variables have a different impact for the individual candidates. For some the Wikipedia traffic was quite unimportant, which was also the case with many other variables.

Table 5.3 Correlation matrix of the collected variables

Variable	1	2	3	4	5	6	7	8	9	10
1										
2	-0.03									
3	-0.02	0.07								
4	-0.01	0.53***	0.09*							
5	0.06	0.45***	0.16***	0.37***						
6	-0.03	0.66***	0.05	0.40***	0.52***					
7	0.03	0.55***	0.04	0.46***	0.56***	0.49***				
8	-0.01	0.75***	0.18***	0.60***	0.55***	0.58***	0.55***			
9	0.11**	0.06	0.11**	0.14***	-0.24***	-0.11**	-0.12**	-0.01		
10	0.07	0.00	0.03	-0.01	0.03	0.02	0.10*	0.07	-0.15***	
11	0.12**	-0.01	-0.05	0.06	0.06	-0.02	0.03	0.01	-0.10*	0.02

Sentiment variables were excluded from the correlation table, because the variables only correlated in statistically significant level to each other and not with any other variable

1 Polls, *2* Tweets, *3* Twitter Followers, *4* Wikipedia Traffic, *5* TV Mentions, *6* Celebrity Tweets, *7* Article Mentions, *8* Google Trends, *9* Facebook Page Likes, *10* Talking About This, *11* Betting

$*p<0.05$; $**p<0.05$; $***p<0.01$

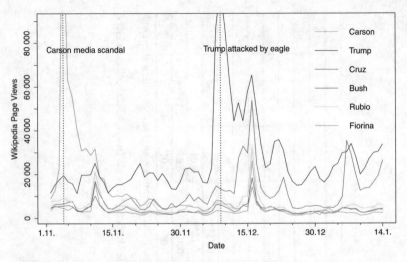

Fig. 5.1 Wikipedia page traffic for the Republican Party candidates (Stats.grok.se 2016)

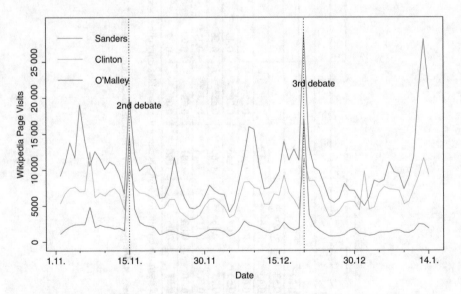

Fig. 5.2 Wikipedia page traffic for the Democratic Party candidates (Stats.grok.se 2016)

The candidates seemed to have quite different relationships to the variables; especially for Hillary Clinton it did not seem to matter that her Wikipedia page views were on a lower level than Bernie Sanders' views in terms of poll ratings. Even Martin O'Malley had more views on debate days. For Sanders, Wikipedia page views had more importance, which was expected because he is not as well known a figure in the presidential race as Clinton; thus, these findings were in line with Yasseri and Bright's (2016) results.

5.4   Discussion and Conclusion

The results of this chapter suggest that machine learning models with linear regression can produce predictions with meaningful accuracy, thus answering to the first research question. Regarding answering the second research question, statistically significant ($p < 0.01$) correlations were found between polls and betting odds and polls and Facebook page likes.

In contrast to what was expected to be found, tweets were not significant in explaining poll results for all candidates, thus the relation of candidates' support to tweets remains in many ways ambiguous. Findings also raise questions about the usefulness of the Twitter Streaming API. It is known to provide only a variable fraction of all the tweets with selected search terms and not necessarily reflecting the real changes in tweet numbers (Morstatter et al. 2013). Besides possible big flaws in the Twitter data itself, the interpretation of tweet numbers is not an easy task. Multiple positive and negative events might be simultaneously affecting candidates' support, both capable of capturing public attention and increasing the candidate's share of tweets.

There would have been an option to combine tweets and sentiment, but that was not pursued, because if tweets were to be combined with sentiment it would be meaningful to combine them to sentiment compared to other candidates or a candidate's own previous sentiment values. This would have been a complex task, yet doable.

We were also thinking of using Twitter user location to identify tweets for individual states, but not everyone uses geolocation and they might not present their user location truthfully, so we gave up the idea. These issues have also been noted in literature (Metaxas et al. 2011).

We noticed in a late stage of data collection that the track-function was only collecting manual retweets, excluding retweets done simply by pressing the retweet button. A candidate's own tweets that do not mention any of the keywords are also excluded. Assigning of the tweet for each candidate could also be improved by combining stronger (e.g. Cruz) and weaker (#GOP) linking words for the candidate, for instance not allocating all tweets that contain word "Cruz" for Ted Cruz, but if "Cruz" and "#GOP" exist in the same tweet it can be counted in. This would increase the accuracy in assigning the tweets for the right candidates.

Twitter verifies accounts that belong to "key individuals and brands", which act in "music, acting, fashion, government, politics, religion, journalism, media, sports, business and other key interest areas" (Twitter Help Center 2016). Being public figures, a candidate's own tweets were included in the celebrity variables. Our results did not show significance of celebrity endorsement in Twitter to the polls.

Betting data has been a good way of predicting results of elections (Wolfers and Leigh 2002), and our findings indicate that it might be worth exploring it further in predicting polls. An interesting idea for future research would be to use betting ratios as a prediction variable. Betting as a variable could prove to be more successful because it reduces noise in the data, which is inherently high in polls.

Predicting elections or polls from social media has proven to be a difficult task, but worth investigating. Future work will extend the list of exploratory variables. Also qualitative analysis of social media content should be included. The increasing public information (e.g., APIs) about individual behavior becomes a potential source for insights on political behavior in the twenty first century.

References

An J, Weber I (2015) Whom should we sense in "social sensing"-analyzing which users work best for social media now-casting. EPJ Data Sci 4(1):1–22

Analysis.gdeltproject.org (2016) http://analysis.gdeltproject.org/module-gkg-exporter.html. Accessed 5 Feb 2016

Berg J, Forsythe R, Nelson F et al (2001) Results from a dozen years of election futures markets research. In: Plott CR, Smith VL (eds) Handbook of experimental economic results, vol 1. Elsevier, Amsterdam, pp 742–751

Berthold MR, Cebron N, Dill F, et al (2007) KNIME: The Konstanz Information Miner. In: Studies in classification, data analysis, and knowledge organization (GfKL 2007), Springer, Heidelberg

Daisey M (2015) Ben Carson's lies reveal a fundamental truth about candidates' tall tales. In: The Guardian. http://www.theguardian.com/commentisfree/2015/nov/14/ben-carsons-lies-reveal-a-fundamental-truth-about-candidates-tall-tales. Accessed 13 Feb 2016

Dalsgaard S (2008) Facework on Facebook: The presentation of self in virtual life and its role in the US elections. Anthropol Today 24(6):8–12

Dalton RJ, Beck PA, Huckfeldt R et al (1998) A test of media-centered agenda setting: Newspaper content and public interests in a presidential election. Polit Commun 15(4):463–481

Democrats.org (2016) The primary debate schedule. https://www.democrats.org/more/the-2016-primary-debate-schedule. Accessed 22 Feb 2016

Dev.twitter.com (2016) https://dev.twitter.com/streaming/overview. Accessed 6 Feb 2016

Erikson RS, Wlezien C (2008) Are political markets really superior to polls as election predictors? Public Opin Q 72(2):190–215

Fox JR, Koloen G, Sahin V (2007) No joke: A comparison of substance in the daily show with Jon Stewart and broadcast network television coverage of the 2004 presidential election campaign. J Broadcast Electron Media 51(2):213–227

Google Trends (2016) https://www.google.com/trends/. Accessed 3 Feb 2016

Griffin A (2015) Donald Trump wants to ban the Internet, will ask Bill Gates to 'close it up'. In: The independent. http://www.independent.co.uk/news/people/donald-trump-wants-to-ban-the-internet-will-ask-bill-gates-to-close-it-up-a6764396.html. Accessed 22 Feb 2016

Guff S (2015) Watch Donald Trump get attacked by a Bald Eagle. In: The Huffington post. http://www.huffingtonpost.com/entry/bald-eagle-shows-trump-whos-boss_us_566881a3e4b0f290e5219f3c. Accessed 22 Feb 2016

Johnson J, Weigel D (2015) Donald Trump calls for 'total' ban on Muslims entering United States. In: Washington post. https://www.washingtonpost.com/politics/2015/12/07/e56266f6-9d2b-11e5-8728-1af6af208198_story.html. Accessed 10 Feb 2016

McKelvey RD, Ordeshook PC (1985) Sequential elections with limited information. Am J Polit Sci 29(3):480–512

McKelvey RD, Ordeshook PC (1986) Information, electoral equilibria, and the democratic ideal. J Polit 48(04):909–937

Metaxas PT, Mustafaraj E, Gayo-Avello D (2011) How (not) to predict elections. Paper presented at the Privacy, Security, Risk and Trust (PASSAT) and IEEE Third International Conference on Social Computing, 9–11 Oct 2011

Morstatter F, Pfeffer J, Liu H et al (2013) Is the sample good enough? Comparing data from Twitter's Streaming API with Twitter's Firehose. arXiv preprint arXiv:1306.5204

Oddschecker.com (2016) http://www.oddschecker.com/politics/us-politics/. Accessed 5 Feb 2016

Pew Research Center (2015) The evolving role of news on Twitter and Facebook. Pew Research Center, Washington

Presidential-candidates.insidegov.com (2016) http://presidential-candidates.insidegov.com/. Accessed 4 Feb 2016

Realclearpolitics.com (2016) http://www.realclearpolitics.com/. Accessed 17 Feb 2016

Sinclair B, Plott CR (2012) From uninformed to informed choices: Voters, pre-election polls and updating. Electoral Stud 31(1):83–95

Stanfordnlp.github.io (2016) http://stanfordnlp.github.io/CoreNLP/. Accessed 11 Feb 2016

Stats.grok.se (2016) http://Stats.grok.se. Accessed 2 Feb 2016

Television.gdeltproject.org (2016) http://television.gdeltproject.org/cgi-bin/iatv_campaign2016/iatv_campaign2016. Accessed 10 Feb 2016

Twitter Help Center (2016) FAQs about verified accounts. https://support.twitter.com/articles/119135. Accessed 21 Feb 2016

Twittercounter (2016) http://twittercounter.com/. Accessed 7 Feb 2016

Vitak J, Zube P, Smock A et al (2011) It's complicated: Facebook users' political participation in the 2008 election. Cyberpsychol Behav Social Networking 14(3):107–114

Wolfers J, Leigh A (2002) Three tools for forecasting federal elections: Lessons from 2001. Aust J Polit Sci 37(2):223–240

Yasseri T, Bright J (2016) Wikipedia traffic data and electoral prediction: towards theoretically informed models. J EPJ Data Sci 22(5):1–15

Zajonc RB (1968) Attitudinal effects of mere exposure. J Pers Soc Psychol 9(2):1–27

Chapter 6
US Election Prediction: A Linguistic Analysis of US Twitter Users

Johannes Bachhuber, Christian Koppeel, Jeronim Morina, Kim Rejström, and David Steinschulte

6.1 Introduction

The world's longest election process started in March 2015 when US Senator Ted Cruz announced his campaign for the Presidential Election of 2016. He was quickly followed by Marco Rubio and Ben Carson. In June 2015, Donald Trump announced his campaign alongside Jeb Bush. On the democratic side, Hillary Clinton was challenged by Bernie Sanders.

Through the popularity of social media and the availability of information on the internet, different forms of social network analysis can provide valuable insights into emerging trends, moods, and values of the public. In addition, information gathering through these sites can easily provide large sample sizes.

In this chapter, we will analyze the linguistic properties of Twitter users to elicit voting preferences based on linguistic tendencies. We will couple this with a focus on the presidential primaries, in an attempt to find a correlation between voter groups and linguistically similar users.

J. Bachhuber
MIT Center for Collective Intelligence, Cambridge, MA, USA

C. Koppeel • J. Morina • D. Steinschulte
University of Cologne, Cologne, Germany

K. Rejström (✉)
Aalto University, Helsinki, Finland
e-mail: kim.rejstrom@aalto.fi

© Springer International Publishing Switzerland 2016
M.P. Zylka et al. (eds.), *Designing Networks for Innovation and Improvisation*,
Springer Proceedings in Complexity, DOI 10.1007/978-3-319-42697-6_6

6.1.1   Research Questions

We are interested in finding distinct groups or clusters of users based on the linguistic properties of their tweets. To do this, we analyzed their word usage for frequencies of certain types of words. Our first research question is

RQ1   Can we identify distinct groups or clusters of users based on natural linguistic analysis of their tweets?

In order to tie this together with the ongoing US Presidential election, we are also interested in investigating if there is a correlation between the users' natural linguistic usage and their voting preferences. Our second research question will address this relation:

RQ2   Can we predict a user's voting preference based on their tweets?

6.2   Prediction Based on Data from Social Media

With the rise of social media websites, users have become increasingly transparent. In addition to the large amount of personal information users provide online, they unwittingly disclose information about their characters, their personality, and even their intelligence through seemingly innocent actions on these platforms. For example, researchers have been able to correlate the *liking* of certain Facebook pages to users' intelligence quotients (Kosinski et al. 2013). Prediction based on data extracted from Social Media is an increasingly growing research area (Asur and Huberman 2010).

Some of the most public, and surprisingly insightful, information users happily disclose online is their use of language. Collecting writing samples from large groups of people used to be difficult, however, with the rise of social media use; the amount of linguistic data available for users has skyrocketed. Users willingly give the entire world access to their thoughts through Facebook, Twitter, Instagram, blogs, and many more sites. For example, as of 2013, Twitter recorded a staggering 500 million tweets per day (Twitter 2013).

The Online Privacy Foundation[1] has conducted research to show just how easy it is to extract information from Facebook and Twitter activity (Sumner et al. 2011; Wald et al. 2012a, b). For example, in one study they collected data on the dark triad of personality traits (narcissism, Machiavellianism, and Psychopathy) from 2927 Twitter users using a Twitter application that asks users a series of questions (Sumner et al. 2012). They then ran a competition on the internet platform Kaggle (Sumner 2012) to find the best algorithms to predict scores of users based on their tweets. The best algorithms reached an accuracy of roughly 86 %. The predictions were not made directly from the tweets, but rather the frequencies of common word groups, which were counted using the Linguistic Inquiry and Word Count (LIWC) software (Pennebaker et al. 2001), within a user's tweets.

---

[1] https://www.onlineprivacyfoundation.org/

Table 6.1 Hashtags and keywords used in Round 1

Candidate	Hashtags
Sanders	#feelthebern,#BernieSanders,#Bernie2016,#UniteBlue,#Sanders2016 ,#Bernie
	@berniesanders,@sensanders
Clinton	#Hillary2016,#trabajocomohillary,#republicanforhillary,#secclinton, #HillaryClinton
	@HillaryClinton
Trump	#makeAmericaGreatAgain,#DonaldTrump,#Trump2016,#Trump,#W ake UpAmerica
	@realdonaldtrump
Rubio	#MarcoRubio,#ImmigrationReform,#Rubio
	@marcorubio,@teammarco
Carson	#bencarson,#BenCarson2016,#BenCarson,#Carson2016,#BC2DC16
	@realbencarson

Tumasjan et al. (2010) used a similar linguistic analysis and tweet frequency approach to show that election prediction could be done based on tweets. They argued that prediction wasn't only doable, it was easy—a statement that has however been disputed several times, most notably by Jungherr et al. (2011).

Although research on predictions based on social media data have rapidly increased, most studies have focused on post hoc analysis of the data (Gayo-Avello 2012). In this chapter, we aim to predict a future result.

6.3   Methodology

6.3.1   Collection of Tweets

In order to collect data from Twitter, we made use of the Twitter Streaming API as well as a custom Voter-Profiler tool. The data gathering and analysis was carried out through three distinct rounds. In the first round, roughly 700,000 relevant tweets were collected through the Twitter Streaming API based on the hashtags and key-words as seen in Table 6.1. In the second round, we collected the full timelines of the manually verified supporters from Round 1 using the Voter-Profiler tool. The total number of Twitter timelines collected in this round was 305 as seen in Table 6.2. These users formed the training dataset for our machine-learning algorithms. In the third round, we used the Twitter Streaming API to collect a new set of election-related tweets from users explicitly stating in their Twitter Profile that they were from the USA. We used self-reported nationality as a simple heuristic for eligibility to vote. We found 2160 users, for whom we then used the Voter-Profiler tool to fetch the users' timelines. The Round 3 users were then used for making the actual prediction. It is worth noting that due to the exploratory nature of this study and rate restrictions of the Twitter API, the size of the datasets used throughout this study are fairly small.

Table 6.2 Number of identified supporters

	Sanders	Clinton	Trump	Rubio	Carson
Supporters	85	51	75	45	49

6.3.2 Analysis of Tweets

Table 6.2 shows the results of the Round 1 collection where we manually identified actual supporters from the collected tweets. This was accomplished by sorting through the 700,000 tweets collected and grouping the Twitter users based on which candidate they state that they support. The target was to identify approximately 50 supporters per candidate for further analysis in the following phases.

We then carried out linguistic analysis using our Voter-Profiler tool on these Twitter users. The Voter-Profiler tool draws from Pennebaker et al. (2001) LIWC research and performs simple frequency analysis of basic word groups. The profiling is done based on frequencies of the following word types: *Negations, Articles, Auxiliary verbs, Prepositions, Conjunctions, Quantifiers, and Pronouns (First singular, Second singular, Third singular, First plural, Second plural, Third plural).*

Using the tool, we were able to create clusters from the supporter groups based on their language usage. The results are discussed further in Sect. 6.4.

6.3.3 Machine Learning

Our initial approach to the machine learning was to use the clusters created by manually identifying supporters and using the frequencies of words groups, as determined by the Voter-Profiler, as training data. The goal was to classify the Round 3 Twitter users based on these supporter clusters. We split our training set so that 182 users were used by the learner and the remaining 123 were used to test the allocations. Unfortunately, a simple prediction with only one learner and predictor with all of our five candidates' identified supporters as input yielded mediocre results. The linguistic differences between the support groups were too small to make reliable predictions, giving us accuracies below 30 % and a Cohen's kappa of 0.069 for the decision tree algorithm.

A reduction of the groups was necessary so that only two possible outcomes remained. The most obvious option was to use Democrats vs. Republicans to determine the party. Using the same data and a decision tree yields an accuracy of approximately 63 % with a Cohen's kappa of 0.27.

As the one learner approach was insufficient, we needed to modify our approach to be a two-step prediction. At first we created two independent predictors. The first distinguishes between *supporter of A and not a supporter of A* and the second between *supporter of B and not a supporter of B*. Both predictors get the same input for prediction. The second step is a necessary comparison of the predicted groups. We thus have four different cases: *"A and not B," "not A and B," "not A and not B,"* and *"A and B."* The first group supports A, the second B, and the third supports a candidate which is not A or B. The fourth group was predicted to support A and B and needed further examination. We set up a third predictor to distinguish between *A and B* for the fourth group.

The approach described above was run with five different learners in KNIME to evaluate the best suited ML algorithm for our task. In most cases, the prediction with a *decision tree* had the highest correctness of allocation. This was the learner we chose to use in the study. The allocations with a *naïve bayes learner* were nearly as good as with a decision tree. We also tried *logistic regression*, *random forest*, and *the SOTA learner*, but those learners allocated only about 30 % of the training data correctly.

6.4   Results

Returning to research question 1 (RQ1), we used the identified supporter groups from Round 1 as a categorization basis for the linguistic analysis. Using the Voter-profiler tool, we extracted language data based on natural linguistic analysis of the users' tweets. This way we were able to create a distinct language profile for each of our five supporter groups.

Research question 2 (RQ2) posed greater challenges, as the language profiles for the supporter groups had very small differences causing the profiles to somewhat overlap. This was also the reason why the initial machine-learning approach proved not feasible.

6.4.1   Linguistic Analysis of the Supporters' Tweets

The supporter groups were analyzed by the Voter-Profiler based on the 13 different word types presented in Sect. 6.3.2. The resulting language profiles are presented in Fig. 6.1. Overall the variations between the language profiles for the support groups were relatively small. There were however some significant differences as

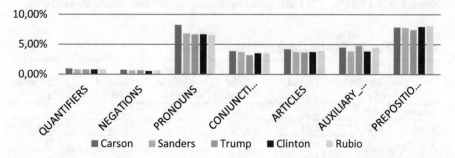

Fig. 6.1 Language profiles grouped by word frequency

well. As seen in Fig. 6.1, the largest differences can be found in Pronoun and Auxiliary Verb usage. Carson's supporter group received a significantly higher usage of pronouns than the supporter groups of the other candidates.

As shown in Table 6.3, the clearest similarity was that all supporter groups used more singular pronouns than plural pronouns. Clinton and Rubio supporters showed almost the same results in every type of pronouns. Carson, Sanders, Clinton, and Rubio showed a high similarity in usage of first person plural pronouns.

There are a few spikes visible, most notably the large percentage of first person singular for Carson's supporters in contrast to Trump's supporters to exhibit the highest first and third person plural usages.

6.4.2 Machine Learning

The automated analysis of the collected tweets took place in three cases. The first one differentiates between the two parties and the other two differentiate between the candidates of each party. The run produced the results shown in Fig. 6.2.

In the first case, the machine-learning algorithm allocated the Twitter users to a distinct party. The user was identified either as a supporter of a democratic candidate or a republican candidate. The voter share of democrats was 55.41%, and the voter share of republicans was 44.59%. According to these results, the next US president would most likely be a member of the Democratic Party.

The next two cases determine the most likely nominee for both parties. In the Democratic Party, we differentiated between supporters of Clinton, Sanders, and another candidate. The algorithm allocated 10.11% of the Twitter users to the supporter group of Clinton and 19.40% to Sanders. According to these results, the nomination of Democratic Party will be won by Sanders.

The considered candidates of the Republican Party are Trump, Carson, and Rubio. To use the same approach used for the democratic candidates, the three republican candidates have to be clustered to two groups. As the results show, Carson and Rubio supporters are more similar than Trump voters. Thus, Carson and Rubio were put together in one option against Trump. The results of the algorithm with these options were that Trump's voter share was 19.68%, while Carson's and Rubio's was 17.91%. The most likely nominated republican candidate is Trump.

Table 6.3 Amount of pronouns used by the supporter groups

	First singular (%)	Second singular (%)	Third singular (%)	First plural (%)	Second plural (%)	Third plural (%)
Sanders	1.82	1.26	1.21	0.54	1.23	0.40
Clinton	1.50	1.14	1.08	0.52	1.14	0.31
Trump	1.12	1.13	1.51	0.76	1.13	0.47
Rubio	1.58	1.05	1.27	0.58	1.05	0.38
Carson	2.43	1.79	1.20	0.64	1.74	0.36

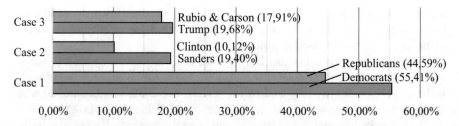

Fig. 6.2 Machine-learning results

6.5   Discussion and Limitations

The aim of the study was to explore the feasibility of election prediction based exclusively on natural language usage in tweets. As seen in Sect. 6.4, we were unable to attain language profiles that were distinctive enough to be able to do accurate categorization through machine learning. Due to the similar language profiles of the supporter groups, it was difficult for the machine-learning algorithms to categorize the Twitter user correctly. One reason for this might be the Twitter APIs. The Streaming API allows tweets to be gathered in real time, but rate restrictions apply when fetching tweets retroactively. These limits make it difficult to achieve large sample sizes for a sufficiently large number of Twitter users. Therefore, the data for the linguistic analysis and training set is based on the timelines of only 305 Twitter users found from Round 2, and the predictions are based on the timelines of 2160 Twitter users from Round 3.

Further, Twitter allows only 140 characters per tweet, which leads to increased use of abbreviations and emoticons to efficiently exploit the limited space. We only included full words that were spelled correctly in our analysis.

As gathering training data for the machine-learning algorithm was very tedious, the training set sizes were very small. We only included Twitter users who explicitly stated that they supported a certain candidate, which was achieved by manually looking through more than 700,000 tweets. Despite this large amount of tweets, only a small set of supporters (45–85) per candidate could be identified.

We also did not account for the selection bias associated with gathering data only on Twitter, so our results are specific to Twitter's demographics. For instance, it is well established that Bernie Sanders is doing exceptionally well among millennials, who are also overrepresented among Twitter's user base (Duggan et al. 2015).

6.6   Future Work and Possible Extensions

Combining the linguistic features studied here with several other parameters, such as sentiment of tweets, number of tweets per candidate, betting site odds, and Google Trends, could greatly improve prediction accuracies. We found that relying solely on linguistic analysis may not be enough. Other parameters were not included in this exploratory study as the goal was to test the feasibility of language analysis alone, in the context of election prediction.

There have been several studies linking word usage to personality traits (Sumner et al. 2012; Sumner 2012; Pennebaker et al. 2001; Gayo-Avello 2012) and other studies linking personality to voting tendencies (Verhulst et al. 2012; Mondak and Halperin 2008), and we feel further research into this area could yield valuable insights.

The Voter-Profiler tool would benefit from several extensions. We currently only support fetching data directly from Twitter, which could be extended to allow importing data directly from databases or other sources. This enables the analysis of tweets bypassing the limitations of the Twitter API. As the tweets in the database are single tweets and not user-timelines, a search and aggregation function should be implemented. This would enable finding tweets based on username.

Furthermore, extending the linguistic capabilities to abbreviations, "Twitter-lingo," and common phrases used in text-based communication would greatly improve the value extracted from each tweet. Emojis could also be included as a base for sentiment calculation, as shown in (Novak et al. 2015).

During the selection of tweets, measures could be introduced to attempt to control for the selection bias discussed in the limitations.

6.7 Conclusion

This chapter is an exploratory study that assesses the feasibility of natural language usage of Twitter users as a base for categorizing political opinion and thus election prediction. The study takes a novel approach to categorize potential voters into groups based on linguistic properties derived from their tweets. A similar approach has previously used linguistic properties to predict personality traits.

The Voter-Profiler tool we introduced can perform basic linguistic analysis on an arbitrary list of Twitter users. The accuracy we achieved through our machine-learning analysis was better than random guessing, but lacks the desired statistical significance. The best accuracies (approx. 63 %) were achieved for the binary categorization into republicans and democrats, whereas categorizations with more candidates are increasingly difficult.

This highlights the necessity for further research in this interesting field.

References

Asur S, Huberman BA (2010) Predicting the future with social media. In: Proceedings of the 2010 IEEE/WIC/ACM International Conference on Web Intelligence and Intelligent Agent Technology, vol 1, IEEE Computer Society, Washington, pp 492–499

Duggan M, Ellison NB, Lampe C, Lenhart A, Madden M (2015) Demographics of key social networking platforms. http://www.pewinternet.org/2015/01/09/demographics-of-key-social-networking-platforms-2/. Accessed 28 Feb 2016

Gayo-Avello D (2012) I wanted to predict elections with twitter and all I got was this Lousy Paper—A balanced survey on election prediction using twitter data. *arXiv preprint arXiv:1204.6441*

Jungherr A, Jürgens P, Schoen H (2011) Why the pirate party won the German election of 2009 or the trouble with predictions: A response to Tumasjan, A., Sprenger, T. O., Sander, P. G., & Welpe, I. M. "Predicting Elections with Twitter: What 140 characters reveal about political sentiment". Social Sci Comput Rev 30(2):229–234

Kosinski M, Stillwell D, Graepel T (2013) Private traits and attributes are predictable from digital records of human behavior. Proc Natl Acad Sci U S A 110:5802–5805

Mondak JJ, Halperin KD (2008) A framework for the study of personality and political behaviour. Br J Polit Sci 38:335–362

Novak PK, Smailović J, Sluban B, Mozetič I (2015) Sentiment of emojis. PLoS One 10, e0144296

Pennebaker JW, Francis ME, Booth RJ (2001) Linguistic inquiry and word count: LIWC 2001. Lawrence Erlbaum Associates, Mahway, p 71

Sumner C (2012) Personality prediction based on twitter stream. https://www.kaggle.com/c/twitter-personality-prediction. Accessed 1 Feb 2016

Sumner C, Byers A, Shearing M (2011) Determining personality traits & privacy concerns from Facebook activity. Black Hat Briefings 11:197–221

Sumner C, Byers A, Boochever R, Park GJ (2012) Predicting dark triad personality traits from twitter usage and a linguistic analysis of tweets. In: Proceedings of the 2012 11th International Conference on Machine Learning and Applications (ICMLA), vol 2, IEEE Computer Society, Washington, pp 386–393

Tumasjan A, Sprenger TO, Sandner PG, Welpe IM (2010) What 140 characters reveal about political sentiment. In: Proceedings of the Fourth International AAAI Conference on Weblogs and Social Media, AAAI Press, Menlo Park

Twitter Inc (2013) New Tweets per second record, and how!. https://blog.twitter.com/2013/new-tweets-per-second-record-and-how. Accessed 14 Feb 2016

Verhulst B, Eaves LJ, Hatemi PK (2012) Correlation not causation: The relationship between personality traits and political ideologies. Am J Polit Sci 56:34–51

Wald R, Khoshgoftaar T, Sumner C (2012a) Machine prediction of personality from Facebook profiles. In: Proceedings of the 2012 IEEE 13th International Conference on Information Reuse and Integration (IRI), IEEE Computer Society, Washington, pp 109–115

Wald R, Khoshgoftaar TM, Napolitano A, Sumner C (2012b) Using twitter content to predict psychopathy. In: Proceedings of the 2012 11th International Conference on Machine Learning and Applications (ICMLA), vol 2, IEEE Computer Society, Washington, pp 394–401

Chapter 7
"Only Say Something When You Have Something to Say": Identifying Creatives Through Their Communication Patterns

Peter A. Gloor, Hauke Fuehres, and Kai Fischbach

7.1 Introduction

In his studies and interviews with famously creative people such as Nobel Prize winners and artists, Csikszentmihalyi (1997) found that these highly creative people share some contradictory traits. They have high physical energy, but are also frequently at rest. They are smart, but also naïve. They combine playfulness with discipline, fantasy with reality, extroversion with introversion, and show both feminine and masculine traits. They are both humble and proud at the same time, independent but also rebellious, and passionate and objective.

In this paper, the creativity of employees in a research and development department of a globally active energy company is analyzed. Characteristics of patterns in communication behavior measured through e-mail are associated with creativity of teams and individuals in organizations (Kidane and Gloor 2007). The e-mail communication of the employees is studied to identify communication patterns suggesting creativity and innovational strength of the individuals. Note that while some authors distinguish between creativity and innovativeness (Margins and Terblanche 2003) we treat the two concepts as interchangeable.

P.A. Gloor (✉)
MIT Center for Collective Intelligence, Cambridge, MA, USA
e-mail: pgloor@mit.edu

H. Fuehres • K. Fischbach
Department of Information System & Social Networks,
University of Bamberg, Bamberg, Germany
e-mail: hauke.fuehres@uni-bamberg.de; kai.fischbach@uni-bamberg.de

© Springer International Publishing Switzerland 2016
M.P. Zylka et al. (eds.), *Designing Networks for Innovation and Improvisation*,
Springer Proceedings in Complexity, DOI 10.1007/978-3-319-42697-6_7

7.2   Background: How to Measure Creativity

Csikszentmihalyi (1997)) defines creativity as an act, idea, or product that changes an existing domain—which can be anything from cooking to nuclear physics—or that transforms an existing domain into a new one. Researchers have been studying individual and organizational creativity for a long time. However, understanding its key ingredients has been elusive. Amabile (1983, 1996) and Amabile et al. (1996) identify creativity as a combination of expertise, creative-thinking skills, and motivation. Expertise consists of procedural, technical, and intellectual knowledge. Intrinsic motivation and inner passion to solve the problem are essential for truly creative solutions, as are creative-thinking skills that include persistence in the face of adversity.

Researchers have identified individual-level characteristics of creativity and the observable influence of these characteristics on output creativity (Helson 1996). However, the influence of the environment on the creative output of the individual has soon been recognized (Perry-Smith 2006; Perry-Smith and Shalley 2003). It has been understood that a combination of skills, motivation, personality, and contextual factors will influence creativity (Woodman et al. 1993; Zhou 2003). Social network analysis techniques have been used to better understand the influence of collaboration on extraordinary creativity (Leenders et al. 2003; Sawyer 2007).

Gloor (2006) postulates that better communication inside an organization will lead to better collaboration, which in turn will lead to better innovation. In follow on work (Gloor et al. 2012), communication among team members has been measured by tracking interpersonal communication through e-mail archives, phone logs, and sociometric badges—body-worn sensors. These communication patterns have then been compared against individual and team creativity. In this paper, an extension of this approach within a global high-tech energy firm is described. Economic growth drives energy demand. An essential global challenge is how the world can continue to grow its global economy, to increase the standard of living of billions of people while minimizing the environmental impact related to the use of energy and economic development. Technology has the highest impact, but also has highest uncertainty. This asks for unparalleled creativity in research and development. Understanding the communication patterns of particularly creative R&D members will help understand and increase our capability to take on the toughest energy challenges.

7.3   Method

In this paper, the corporate e-mail network of the thousands of employees of the research and development department of a global energy company is analyzed. The e-mail traffic among those thousands of employees has been collected over 13 months. Their e-mail communication with outside researchers from universities has also been included in the dataset. Only the structural data on the e-mail communication were analyzed. Neither content information nor e-mail subject lines were evaluated. The information on the senders and receiver for e-mails has been anonymized.

In order to test the assumptions made in this paper, the creativity and innovative strength of the employees is measured. Two classes of additional characteristics on the employees have thus been collected: Whether an employee filed for patents or published scientific papers shows output-oriented behavior. As internal, outcome-oriented behavior two internal awards have been taken into account. The first award rewards the most exceptional patent filed by an employee, the second award is granted to the most innovative employee.

For this study, creativity and innovational strength of an employee is identified by four measures:

Output metrics:

Number of Patents $_{Emp}$ = Number of patents filed by employees
Number of Best Papers $_{Emp}$ = Number of "publications of the year" by employee

Outcome metrics:

Number of Edison Awards $_{Emp}$ = Number of Thomas Alva Edison Patent Award (honoring the most exceptional efforts of scientists and inventors)
Number of Most Innovative Employee $_{Emp}$ = Number of annual awards for the most innovative employee rewarded to employee

We call employees who hold at least one of those four criteria innovators. In addition, innovators have been categorized into *output* innovators—filing for patents and writing best papers, and *outcome* innovators—getting an award. In an additional analysis, the difference in communication patterns between *repeat* innovators—who fulfill more than one criterion or one criterion several times—and innovators who fulfill a criterion once—we call them "*one-shot* innovators"—has been studied.

7.4 Results

We constructed the full network of the thousands of members of the R&D department. For each member of the department we calculated their six honest signals of e-mail collaboration, as defined in Gloor (2015). They are shown in Table 7.1. Figure 7.1 shows the in-group network of the thousands of members of the department for 2015, colored by divisions. Each dot is an actor, each connecting line means that at least one e-mail has been exchanged between the two actors.

7.4.1 Innovators Among Their Peers

When looking at the network structure and the six honest signals of communication, innovators stand out in several ways from their peers in the analyzed department. The innovators were more respected, as their peers within the department answered them on average in 20 h instead of the 22 h it took for everybody else to get at response (Table 7.2).

Table 7.1 Six honest signals of collaboration

Indicator	SNA term	Definition	How the variables are calculated in condor
Central leadership	Degree centrality	Number of actors each person is directly connected within a network	Is the number of nearest neighbors from an actor both as senders or receivers in the network
	Betweenness centrality	It is a measure of the extent to which each actor acts as an information hub and controls the information flow	It is defined as the likelihood to be on the shortest path between any two actors in the network
Rotating leadership	Betweenness centrality oscillation	It is a measure of how frequently actors change their network position in the team, from central to peripheral, and back	Number of local maxima and minima in the betweenness curve of an actor or a group
Balanced contribution	Contribution index	Indicates how balanced a communication is in terms of msg sent and msg received	Msg sent-msg rcvd/(msg sent+msg rcvd)
Rapid response	Ego ART	Average number of hours sender takes to respond to e-mails	Time until a frame is closed for the receiver after he has sent an e-mail
	Ego nudges	Average number of follow-ups that the sender needs to send in order to receive a response from the receiver	Number of pings until sender responds
	Alter ART	Average number of hours receiver takes to respond to e-mails	Time until a frame is closed for the sender, after he has sent an e-mail
	Alter nudges	Average number of follow-ups that the receiver needs to send in order to receive a response from the sender	Number of pings until receiver responds
Honest language	Avg. sentiment	Indicates positivity and negativity of communication	Uses automatically generated bag of word, based on a dictionary trained for language/subject area
	Avg. emotionality	Represents the deviation from neutral sentiment	Standard deviation of sentiment
Shared context	Avg. complexity	It is a measure of complexity of word usage. It is defined as the information distribution, i.e. the more diverse words, which are all used evenly, a sender uses, the higher his complexity	Information distribution using TF/IDF, independent of single words

Fig. 7.1 Social network of the R&D department, colored by divisions

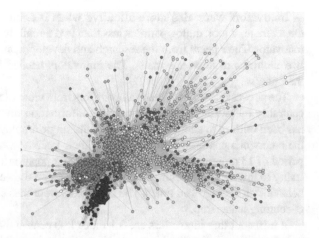

Table 7.2 Social network metrics comparing innovators with their peers (**significant on the 0.01 level, *significant on the 0.05 level)

		N	Mean	Std. Deviation
Messages sent	Peers	1718	6760.55	7969.23
	Innovators	194	6265.12	5669.05
Alter ART [h]**	**Peers**	**1718**	**21.93**	**7.80**
	Innovators	**194**	**20.37**	**6.47**
Messages received	Peers	1718	6270.62	3773.43
	Innovators	194	6756.89	3956.28
Messages total	Peers	1718	13031.17	10702.61
	Innovators	194	13022.01	9156.42
Contribution index**	**Peers**	**1718**	**−0.09**	**0.27**
	Innovators	**194**	**−0.14**	**0.24**
Ego ART [h]	Peers	1718	19.85	5.60
	Innovators	194	19.29	4.32
Ego nudges**	**Peers**	**1718**	**2.01**	**0.44**
	Innovators	**194**	**1.83**	**0.31**
Betweenness centrality oscillation*	**Peers**	**1718**	**166.26**	**27.85**
	Innovators	**194**	**170.04**	**21.60**
Alter nudges**	**Peers**	**1718**	**2.09**	**0.40**
	Innovators	**194**	**2.00**	**0.29**
Betweenness centrality**	**Peers**	**1718**	**5302.00**	**43651.09**
	Innovators	**194**	**1707.57**	**3568.16**
Degree centrality	Peers	1718	260.68	196.02
	Innovators	194	274.86	163.69
Contribution index oscillation	Peers	1718	164.81	29.87
	Innovators	194	167.35	22.10

Innovators were also more attentive when it comes to answering e-mails. On average, it took dialog partners less than two e-mails to elicit an answer from an innovator. Their peers from the research and development department needed over two inquiries to answer e-mails. The innovators needed 5 % less reminders than their peers.

The network position of innovators differs from those of their peers. Betweenness centrality measuring the control over the information flows in the e-mail network has been utilized to show the influence of the single employees. Innovators changed their network position more often than their peers. Over the entire observation period of 13 months, they changed their network position 170 times from high centrality to low centrality, compared to their peers who showed the rotating leadership behavior on average 166 times. The innovators were thus showing a higher amount of rotating leadership.

It turns out that innovators are 3.1 times less central than their peers, measured in betweenness centrality. They are also more passive senders of e-mail in terms of their contribution of new messages. When restricting the communication to the local lab of the innovators, however, they switch roles, and become more central within their local lab network than their peers. The innovators are then 2.2 times more central than their direct peers.

7.4.2 Communication with External Researchers

The communication behavior with people outside of the organization also differed between innovators and their peers. Communication of employees with researchers from universities has been analyzed. The innovators got 1.7 times more e-mails from senders with university e-mail addresses, and they were sending 1.4 times more e-mail to university researchers compared to their peers within the R&D department. Their rotating leadership behavior became even more pronounced in communication with universities, they changed their position from being the leader to the listeners 65 times in the 13 months observation period, compared to 41 rotational changes in leadership for their peers within the company. Innovators also talked with more different people at the university over these 13 months, initiating an e-mail dialog with 56 outside people instead of 37 (Table 7.3).

7.4.3 Output-Oriented and Outcome-Oriented Innovators

Innovators at the examined department can further be partitioned into two groups of innovators. Output-oriented innovators are those who fulfill the first two criteria of being an innovator: They publish scientific papers and file patents. The second group of innovators, the outcome-oriented innovators, are employees who were

Table 7.3 Social network metrics comparing innovators with their peers when communicating with outside academics peers (**significant on the 0.01 level, *significant on the 0.05 level)

		N	Mean	Std. deviation
Messages sent	Innovators	103	268.78	361.32
	Peers	165	189.79	313.92
Alter ART [h]	Innovators	97	20.20	15.44
	Peers	145	16.85	16.95
Messages received**	**Innovators**	**103**	**129.54**	**113.85**
	Peers	**165**	**77.05**	**109.17**
Messages total**	**Innovators**	**103**	**398.32**	**443.33**
	Peers	**165**	**266.85**	**380.09**
Contribution index*	**Innovators**	**103**	**0.12**	**0.42**
	Peers	**165**	**0.24**	**0.37**
Ego ART [h]	Innovators	103	22.19	19.38
	Peers	165	18.44	17.85
Ego nudges	Innovators	103	1.65	0.48
	Peers	165	1.69	0.65
Betweenness centrality oscillation**	**Innovators**	**103**	**65.17**	**41.25**
	Peers	**165**	**41.92**	**40.39**
Alter nudges	Innovators	97	1.19	0.26
	Peers	145	1.24	0.49
Betweenness centrality*	**Innovators**	**103**	**4.23E+06**	**1.07E+07**
	Peers	**165**	**2.32E+06**	**3.78E+06**
Degree centrality**	**Innovators**	**103**	**55.99**	**54.48**
	Peers	**165**	**37.21**	**39.48**
Contribution index oscillation**	**Innovators**	**103**	**54.61**	**39.22**
	Peers	**165**	**38.75**	**37.51**

rewarded with the Edison patent award or with awards for the most innovative employee. Paper and patent writing innovators show more introvert behavior than their peers, while award winning innovators show more extrovert behavior: while on average innovators send less e-mail than their peers, award winning innovators were the most active senders and receivers of e-mail in the entire R&D department. However, it seems that paper and patent writing innovators are more respected than award winning innovators, as they are responded faster than award winning innovators. However this could also be related to the higher amount of e-mail generated by outcome-oriented innovators, as this puts a higher strain on others to answer them. Paper and patent writing innovators are less central—more introvert—than award winning innovators, who must be highly visible in order to be nominated for their awards. Award winning innovators also show more rotating leadership, switching between leading and listening, than the more introvert paper and patent writing innovators (Table 7.4).

Table 7.4 Social network metrics comparing outcome innovators with output innovators

		Messages sent	Alter ART [h]	Messages received	Messages total	Contribution index	Ego ART [h]	Ego nudges	Betweenness centrality oscillation	Alter nudges	Betweenness centrality	Degree centrality	Contribution index oscillation
Outcome innovators	Mean	7196.47	**20.57**	7640.18	**14836.65**	-0.11	19.04	**1.89**	171.65	**2.03**	2253.23	275.92	167.34
	N	74	**74**	74	**74**	74	74	**74**	74	**74**	74	74	74
	Std. deviation	5924.62	**6.01**	4322.13	**9509.54**	0.23	3.45	**0.31**	23.32	**0.36**	4366.79	145.59	22.73
Output innovators	Mean	5690.78	**20.25**	6212.19	**11902.98**	-0.17	19.44	**1.80**	169.04	**1.98**	1371.08	274.21	167.36
	N	120	**120**	120	**120**	120	120	**120**	120	**120**	120	120	120
	Std. deviation	5451.47	**6.76**	3625.10	**8785.83**	0.25	4.78	**0.30**	20.50	**0.24**	2941.43	174.51	21.80
Peers	Mean	6760.55	**21.93**	6270.62	**13031.17**	-0.09	19.85	**2.01**	166.26	**2.09**	5302.00	260.68	164.81
	N	1718	**1718**	1718	**1718**	1718	1718	**1718**	1718	**1718**	1718	1718	1718
	Std. deviation	7969.23	**7.80**	3773.43	**10702.61**	0.27	5.60	**0.44**	27.85	**0.40**	43651.09	196.02	29.87
Total	Mean	6710.28	**21.77**	6319.96	**13030.24**	-0.10	19.79	**1.99**	166.65	**2.08**	4937.29	262.12	165.07
	N	1912	**1912**	1912	**1912**	1912	1912	**1912**	1912	**1912**	1912	1912	1912
	Std. deviation	7767.21	**7.68**	3794.16	**10553.90**	0.26	5.49	**0.43**	27.30	**0.39**	41405.90	192.99	29.18

Bold denotes a significant difference (non-parametric Kruskal–Wallis test)

7.4.4 Repeat Innovators

Finally, we also looked at whether there is a difference between repeat innovators—who fulfill more than one criterion or one criterion for several times—and innovators who fulfill a criterion one once—we call them the "one-shot innovators". We found that repeat innovators sent twice as much e-mail to their peers. Repeat innovators show higher rotating leadership, changing from leading to listening 30 % more than one-shot innovators. They also command higher respect—it takes less nudges for others to answer back to them, and they are more central in the network, although the difference in betweenness centrality is not statistically significant (Table 7.5).

Table 7.5 Social network metrics comparing repeat innovators with "one-shot" innovators (**significant on the 0.01 level, *significant on the 0.05 level)

		N	Mean	Std. deviation
Messages sent*	**One-shot innov.**	**23**	**118.74**	**131.83**
	Repeat innovators	**78**	**279.03**	**376.96**
Alter ART [h]	One-shot innov.	18	20.27	12.10
	Repeat innovators	70	20.25	13.97
Messages received	One-shot innov.	23	82.96	63.77
	Repeat innovators	78	125.13	122.12
Messages total*	**One-shot innov.**	**23**	**201.70**	**167.16**
	Repeat innovators	**78**	**404.15**	**468.13**
Contribution index	One-shot innov.	23	0.09	0.46
	Repeat innovators	78	0.20	0.35
Ego ART [h]	One-shot innov.	15	18.46	17.25
	Repeat innovators	68	22.30	16.82
Ego nudges	One-shot innov.	15	1.60	0.43
	Repeat innovators	68	1.72	0.49
Betweenness centrality oscillation*	**One-shot innov.**	**23**	**42.78**	**28.90**
	Repeat innovators	**78**	**64.01**	**43.23**
Alter nudges*	**One-shot innov.**	**18**	**1.39**	**0.54**
	Repeat innovators	**70**	**1.19**	**0.27**
Betweenness centrality	One-shot innov.	23	1.96E+06	1.92E+06
	Repeat innovators	78	4.51E+06	1.22E+07
Degree centrality	One-shot innov.	23	32.957	25.677
	Repeat innovators	78	54.96	57.74
Contribution index oscillation*	**One-shot innov.**	**23**	**35.74**	**25.85**
	Repeat innovators	**78**	**53.71**	**40.39**
Newcontact	One-shot innov.	23	101.87	55.92
	Repeat innovators	74	119.36	53.31

7.5 Discussion

In sum, we find that intrinsically motivated innovators can be found by looking at the "honest signals of collaboration", with signs of respect shown as faster response by others, and the innovators showing high passion by answering their e-mails faster than their peers. A second key insight is that innovators are less "political" in their e-mailing behavior by sending much less e-mail within their organization. They use their communication bandwidth to extend their external network to develop new innovative ideas. This helps them minimize organization's internal echo chambers of thought, and constantly add divergent thinking from external entities—thus increasing their "innovation quotient" in the organization. In addition, innovators exhibit significantly higher rotating leadership through oscillations in betweenness centrality.

One limitation our study is the focus on publication and patent productivity, which does not imply correlation with output quality. A more realistic metric would be the financial impact of the patents on the bottom line of the company. Unfortunately, these numbers were not available to us. Another open question is causality, it might be that better connected employees at the company have an easier time to file a patent, in particular because having many patents filed is considered a major means of success for an R&D employee, so it might be that this is an instance of the "rich get richer" syndrome, and once somebody has filed a patent, it will become successively easier to file more patents.

In combination, looking at these honest signals of collaboration gives valuable insights both to individual innovators and their managers. For individuals, it tells them that it pays to focus on their work, and reach out to the outside for novel ideas. It also encourages them to "take time off for thinking" in between intensive information exchanges with their peers. For managers, the lesson is clear: do not reward "political behavior" by spamming others with too many messages, but nurture a self-organizing emergent leadership style. Employees should be encouraged to employ a more intrinsically motivated communication style (Pink 2011), to "only say something when they have something to say". Employees are encouraged to reach out and connect to outside sources of ideas and innovation. They are allowed to form small ad-hoc workgroups, Collaborative Innovation Networks (COINs) (Gloor 2006) to pursue unconventional "crazy ideas".

References

Amabile TM (1983) The social psychology of creativity: a componential conceptualization. J Pers Soc Psychol 45(2):357–376

Amabile TM (1996) Creativity in context: update to the social psychology of creativity. Westview Press, Boulder

Amabile TM, Conti R, Coon H, Lazenby J, Herron M (1996) Assessing the work environment for creativity. Acad Manage J 39(5):1154–1184

Csikszentmihalyi M (1997) Creativity: flow and the psychology of discovery and invention. Harper Perennial, New York

Gloor PA (2006) Swarm creativity, competitive advantage through collaborative innovation networks. Oxford University Press, New York

Gloor PA (2015) What email reveals about your organization. Sloan Management Review, Winter

Gloor PA, Grippa F, Putzke J, Lassenius C, Fuehres H, Fischbach K, Schoder D (2012) Measuring social capital in creative teams through sociometric sensors. Int J Organ Des Eng 2(4):380–401

Helson R (1996) In search of the creative personality. Creat Res J 9:295–306

Kidane YH, Gloor PA (2007) Correlating temporal communication patterns of the eclipse open source community with performance and creativity. Comput Math Organ Theory 13(1):17–27

Leenders RTAJ, van Engelen JML, Kratzer J (2003) Virtuality, communication, and new product team creativity: a social network perspective. J Eng Technol Manag 20:69–92

Margins EC, Terblanche F (2003) Building organizational culture that stimulates creativity and innovation. Eur J Innov Manag 6(1):64–74

Perry-Smith JE (2006) Social yet creative: the role of social relationships in facilitating individual creativity. Acad Manage J 49(1):85–101

Perry-Smith JE, Shalley CE (2003) The social side of creativity: a static and dynamic social network perspective. Acad Manage Rev 28(1):89–106

Pink DH (2011) Drive: the surprising truth about what motivates us. Penguin, New York

Sawyer C (2007) Group genius: the creative power of collaboration. Basic Books, New York

Woodman RW, Sawyer JE, Griffin RW (1993) Toward a theory of organizational creativity. Acad Manage Rev 18(2):293–321

Zhou J (2003) When the presence of creative coworkers is related to creativity: role of supervisor close monitoring, developmental feedback, and creative personality. J Appl Psychol 88:413–422

Chapter 8
Some Insights into the Relevance of Nodes' Characteristics in Complex Network Structures

Matteo Cinelli, Giovanna Ferraro, and Antonio Iovanella

8.1 Introduction

Complex networks consist of numerous nodes and intricate connections embedded with heterogeneous network structure under the graph-theoretic point of view. An important class of networks with this feature is called scale-free (Barabási and Albert 1999) and often reproduces interactions occurring among entities in real systems (Wang and Chen 2003). In Gloor (2006), it is stated that this representation is still valid in case of Collaborative Innovation Networks (COINs), defined as "cyberteams of self-motivated people with a collective vision, enabled by the Web to collaborate in achieving a common goal by sharing ideas, information and work". Scientific literature has paid particular attention to the influence of the structure on innovation networks and some examples can be found in Cowan and Jonard (2004), Choi et al. (2010), and Ferraro and Iovanella (2015, 2016).

Herein, we analyze a further element in scale-free networks: whether a given nodes' characteristic correlates with the network structure. Indeed, we study if the number of links, between nodes sharing a common characteristic, is able to detect the spreading of such feature and whether it follows a random distribution or vice versa, correlating with the network structure (Park and Barabási 2007). In other terms, we consider if the growth of scale-free networks is due not only to the increase of the nodes and to the preferential attachment mechanism (Barabási and Albert 1999) but also to the homophily, i.e. the tendency of nodes to connect with others similar to themselves (de Almeida et al. 2013). Such phenomenon is called *dyadic effect*. Networks homophily can be characterized on a finite space by two measures called *dyadicity* and *heterophilicity* indicating how similar or dissimilar

M. Cinelli • G. Ferraro • A. Iovanella (✉)
Department of Enterprise Engineering, University of Rome Tor Vergata,
Via del Politecnico, 1, Rome 00133, Italy
e-mail: matteo.cinelli@uniroma2.it; giovanna.ferraro@uniroma2.it;
antonio.iovanella@uniroma2.it

© Springer International Publishing Switzerland 2016
M.P. Zylka et al. (eds.), *Designing Networks for Innovation and Improvisation*,
Springer Proceedings in Complexity, DOI 10.1007/978-3-319-42697-6_8

nodes tend to connect among themselves. Such measures can be analyzed with the approach presented in Park and Barabási (2007) and extended in Ferraro et al. (2016). Nevertheless, despite their usefulness, these approaches are limited by the computational complexity that increases exponentially with the size of the network.

Homophily has received particular attention by the scientific literature under many points of view. For instance, Aral et al. (2009) investigated homophily-driven diffusion in a dynamic network, while Bianconi et al. (2009) defined a general indicator that quantifies how much the topology of a network depends on a given assignment of nodes' characteristic. In real systems, several exogenous nodes' characteristics can be considered. For example, in Hu et al. (2015) the authors face 808 different characteristics in order to study dyadicity and heterophilicity of gene–gene network interactions in cancer research. Thus, it becomes essential to understand which feature is relevant in taking part in the probability that one node links to others. Another approach is in Ferraro et al. (2016) where authors show that the Global Innovation Index correlates with the structure of the Enterprise Europe Network, a technology transfer network founded by the European Commission.

The aim of this paper is to introduce a methodology that, through dealing efficiently with complexity, allows us to give some preliminary hints about the uniqueness of the distribution of nodes' features and its correlation with network architecture. Since we make use of the implementation of very simple formulas, we can handle networks regardless of their size. Thus, even large network can be studied with a minimum effort.

The paper is organized as follows: Sect. 8.2 surveys the theoretical background, Sect. 8.3 outlines the methodology, and Sect. 8.4 presents the conclusions.

8.2 Theoretical Background

The classical mathematical abstraction of a network is a graph G. A graph $G = (V, E)$ is composed of a set V of N nodes and a set E of M links that defines the relationship among these nodes. We refer to a node by an index i, meaning that we allow a one-to-one correspondence between an index and a node. Park and Barabási (2007) noted that, when nodes in a network fit within two distinct groups according to their characteristics, two different parameters are required to catch the relations between the network topology and nodes' features. In many systems, the number of links between nodes sharing a common characteristic is larger than expected if the characteristics are distributed randomly on the graph; this phenomenon is called the *dyadic effect* (White and Harary 2001).

Herein, we refer to a given exogenous characteristic c_i, which can assume the values 0 or 1, for each i in N. So, N can be divided into two subsets: n_1, the set of nodes with characteristic $c_i = 1$, n_0, the set of nodes with characteristic $c_i = 0$; therefore, $N = n_1 + n_0$. For an *exogenous characteristic*, we consider a node feature that does not derive directly from the network topology, nor is influenced by the network

Fig. 8.1 Types of dyads

itself. A non-binary characteristic can be also considered having the caution to consider under or over-performance through the introduction of a threshold value (Ferraro et al. 2016).

We distinguish three kinds of dyads, links and their two end nodes, in the network: (1–1), (1–0), and (0–0) as depicted in Fig. 8.1.

We label the number of each dyad in the graph as m_{11}, m_{10}, and m_{00}, respectively. Hence, $M = m_{11} + m_{10} + m_{00}$. We consider m_{11} and m_{10} as independent parameters that represent the dyads containing nodes with characteristic 1. If a casual setting among the N nodes is considered, where any node has an equal chance of having the characteristic 1, the values of m_{11} and m_{10} are $\bar{m}_{11} = pn_1(n_1 - 1)/2$ and $\bar{m}_{10} = pn_1(N - n_1)$ (Park and Barabási 2007), where p is the *density* equal to $p = 2M/N(N-1)$ with $0 \leq p \leq 1$. The relevant deviations of m_{11} and m_{10} from the expected values $\bar{m}_{11}$ and $\bar{m}_{10}$ denote that the characteristic 1 is not randomly distributed (Park and Barabási 2007; de Almeida et al. 2013). Such deviations can be calculated through the ratios of dyadicity D and heterophilicity H defined as: $D = m_{11}/\bar{m}_{11}$ and $H = m_{10}/\bar{m}_{10}$.

If the characteristic is *dyadic*, $D > 1$, it means that nodes with the same characteristics tend to link more tightly among themselves than expected in a random configuration. Whereas with $D < 1$ the characteristic is *anti-dyadic*, indicating that similar nodes tend to connect less densely among themselves than expected in a random configuration. The characteristic is defined as *heterophilic*, with a value $H > 1$, indicating that nodes with the same features have more connections to nodes with different characteristics than expected randomly. On the contrary, with a value $H < 1$, the characteristic is defined *heterophobic*, meaning that nodes with certain characteristics have fewer links to nodes with different characteristics than expected randomly. We note that m_{11} and m_{10} cannot assume arbitrary values, as there are indirect constraints due to the network structure. Indeed, m_{11} cannot exceed

$$UBm_{11} = \min\left(M, \binom{n_1}{2}\right)$$ and m_{10} cannot be larger than $UBm_{10} = \min(M, n_1 n_0)$,

where UB stands for upper bound.

One instrument to investigate the correlation among the distribution of a given characteristic c and the underlying network structure is the phase diagram that, in general, describes the admissible configurations in the graph.

We consider, as an example, the graph shown in Fig. 8.2, which depicts a network with 25 nodes and 32 links of which $n_1 = 10$ black nodes are randomly distributed (Park and Barabási 2007). The corresponding phase diagram depicted in Fig. 8.3 describes the distribution of a random feature in the system. It should be noted that the reported phase diagram shows only a subarea.

The phase diagram presents all the admissible combinations of m_{10} (*x*-coordinate) and m_{11} (*y*-coordinate) and each corresponding square collects the number of the assignment of n_1 nodes over the set N for every fixed m_{10} and m_{11}. A direct correspondence

Fig. 8.2 An example of graph with $N=25$ and $M=32$

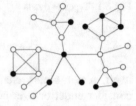

Fig. 8.3 Phase diagram

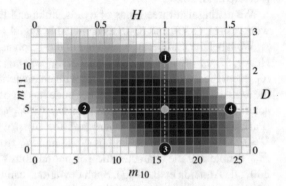

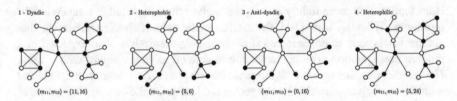

Fig. 8.4 Configurations of four extreme points on the phase diagram

exists among the m_{10} and m_{11} axis and, respectively, H and D, since the values are related by means of the given formulas. Moreover, m_{10} ranges from 0 to UBm_{10} and m_{11} from 0 to UBm_{11}. Correspondingly, D ranges from 0 to $D_{max}=UBm_{11}/\bar{m}_{11}$ and H ranges from 0 to $H_{max}=UBm_{10}/\bar{m}_{10}$. Each square has a darkness proportional to the degeneracy of the configuration, i.e. for a given m_{10} and m_{11}, an open square means that is not possible to place n_1 nodes consistently with fixed values and network topology constraints.

The phase diagram has some meaningful areas to discuss. In particular, open squares represent non-admissible configurations for fixed m_{10} and m_{11}; the high degeneracy squares are considered as the most typical configurations for a random distribution of a characteristic $D=H=1$; and the phase boundaries squares map atypical configurations. For such phase boundaries different layouts are recognizable. Indeed, in Fig. 8.4, four possible configurations are represented (where the point of each configuration is correspondingly numbered in the phase diagram of Fig. 8.3): $D\gg 1$ is a dyadic case where black nodes concentrate in a central cluster of the graph which maximizes m_{11}; $D\ll 1$ is an anti-dyadic configuration where black nodes tend to be farther apart; $H \ll 1$ is an heterophobic configuration where black nodes are located in the peripheral area which minimize m_{10}; and, $H \gg 1$ is an heterophilic configuration where black nodes relate to the hubs so that the connections with open nodes are maximized.

The graphical nature of the phase diagram allows for easier observation of the distribution of the nodes' characteristics. Though, as the number of the possible configurations increases exponentially with N, the phase diagram is hard to compute for large networks. Moreover, it is worth to mention that multiple simultaneous characteristics lead the problem to be even more difficult since the phase diagram becomes multidimensional (Park and Barabási 2007).

8.3   Methodology

For relatively small networks, the relevance of nodes' characteristics can be studied using the approach presented in Ferraro et al. (2016). However, in case of large systems, this is not applicable due to the increase in computational complexity. In this section, we propose a methodology able to infer some findings from the phase diagram and use them as a pre-processing step in order to give a first approximated result on the relevance of a given nodes' characteristic.

The phase diagram has the properties that its shape is size independent (Park and Barabási 2007) and $\bar{m}_{11}$, $\bar{m}_{10}$, UBm_{11} and UBm_{10} can be computed through simple formulas. Therefore, independently from the network size, we can identify and compute the central and the extremal coordinates of the phase diagram feasible region. Thus, we use such properties to restrict our analysis to examples of relatively small networks while we extend the study to different density values, starting from the very simple binary tree to the complete network when $p=1$.

Our methodology originates from the analysis of the sequence of the $N+1$ phase diagrams computed varying n_1 from 0 to N (see Fig. 8.5) for the network in Fig. 8.2. We observe: when $n_1=0$ the phase diagram collapses in a single square in the left lower side of the diagram, having all nodes with characteristic 0; similarly, when $n_1=N$ the phase diagram collapses in a single square in the left upper side of the diagram, having all nodes with characteristic 1; as n_1 grows, the evolution of the phase diagrams suggests the existence of a trajectory in which the diagram runs.

The phase diagrams have different settings. The first setting is characterized by low values of dyadicity and heterophilicity. We notice that the diagrams start to enlarge allowing more admissible configurations. Finally, in the last setting when n_1 assumes high values, the configurations decrease and dyadicity reaches its maximum value while heterophilicity decreases. Looking at the sequence, it is quite clear that the evolving areas of the phase diagram are built around a highly degenerative core, where the number of configurations is exponentially higher with respect to those in the lighter area. Knowing that the central area of the phase diagram represents the neighborhood of the $D=H=1$ point, we determine which configurations are far from the core. Furthermore, the structure of the network

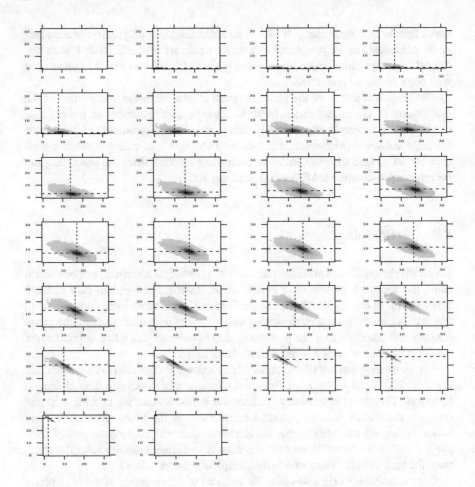

Fig. 8.5 The sequences of the phase diagram of the network depicted in Fig. 8.2 when $n_1 \in [0,N]$, starting from left to right and from the upper to the lower side of the figure. *Dotted lines* indicate the coordinates of $D=H=1$

constraints the available configurations since not all the values of dyadicity and heterophilicity are admissible and cannot vary independently.

In general, if we consider a nodes' characteristic we are interested mainly in how much the current values of dyadicity D and heterophilicity H are far from the area of maximum degeneration. It is easy to compute the values D and H as well as the value of $D=H=1$ for all the values of $n_1 \in [0,N]$ since such points correspond to $\bar{m}_{11}$ and $\bar{m}_{10}$ considering formulas presented in Sect. 8.2.

For the $N+1$ pairs of $\bar{m}_{11}$ and $\bar{m}_{10}$, we can merge such values in a single plot interpolated in a curve. This approach allows distinguishing—with a certain level of approximation given by the relation among the current value of D and H and both the average and the upper bounds values—if the network with the considered characteristics is worth the effort to study without computing the whole phase diagram.

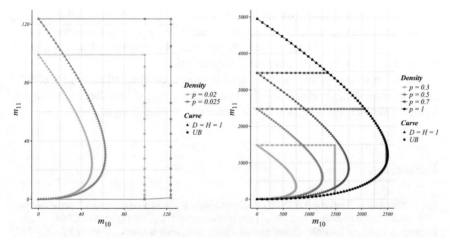

Fig. 8.6 The different curves for the values D and H and upper bounds. The *left figure* shows values of $p=0.02$ and 0.025; the *right figure* shows values of $p=0.3$, 0.5, 0.7, and 1

Indeed, decisions are made in the following way: if the current configuration has the point (D, H) very close to the curve, then it is randomly distributed on the network and does not correlate with the network structure. On the contrary, if the point is located far from the curve, we can continue the analysis of the network referring to, for instance, Park and Barabási (2007), de Almeida et al. (2013) or Ferraro et al. (2016) approaches according to the desired results.

Using the software R and the package *igraph* we built six networks with $N=100$ and diverse density ($p=0.02$ for the binary tree; $p=0.025$; $p=0.3$; $p=0.5$; $p=0.7$; $p=1$ for the clique) in order to study the different curve trends. For each of these networks, we consider a pair of curves (see Fig. 8.6). The first represents the interpolation of the $N+1$ pairs of $\bar{m}_{11}$ and $\bar{m}_{10}$, the second denotes the interpolation of the $N+1$ maximum values calculated by the analytic upper bound which for each n_1 is equal to $UBm_{11} = \min\left(M, \binom{n_1}{2}\right)$ and $UBm_{10} = \min\left(M, n_1 n_0\right)$. If we consider that points D and H are bounded below by a lower bound (LB) equal to the x and y axis, then for a fixed p, the area included from the LB to the UB curve represents the whole domain of the theoretical admissible dyadicity and heterophilicity pairs for every value assumed by n_1.

8.4   Conclusions

In this paper, we developed a new methodology to investigate the correlation between the distribution of nodes' characteristics and the network's topology. In particular, we are able to analyze even large networks without considering the entire phase diagram. By knowing few parameters, such as density and number of featured nodes, we are able to assess if a certain point in the *D-H* space is sufficiently far from the degenerative area, i.e. if a characteristic is randomly distributed along the network. This approach is particularly useful when several characteristics are taken into account in order to identify functional properties of nodes and the computation of every phase diagram is impracticable.

Further research should be conducted to extend the methodology in a broader range of contexts and networks and, in particular, it would be valuable to compute tighter lower and upper bounds that consider combinatorial arguments.

References

Aral S, Muchnik L, Sundarajan A (2009) Distinguishing influence-based contagion from homophily-driven diffusion in dynamic networks. Proc Natl Acad Sci U S A 106(51):21544–21549

Barabási A-L, Albert R (1999) Emergence of scaling in random networks. Science 286:509–512

Bianconi G, Pin B, Marsili M (2009) Assessing the relevance of node features for network structure. Proc Natl Acad Sci U S A 106(28):11433–11438

Choi H, Kim S-H, Lee J (2010) Role of network structure and network effects in diffusion of innovation. Ind Mark Manag 39:170–177

Cowan R, Jonard N (2004) Network structure and the diffusion of knowledge. J Econ Dyn Control 28:1557–1575

de Almeida ML, Mendes GA, Viswanathan GM, da Silva LR (2013) Scale-free homophilic network. Eur Phys J B 86:38

Ferraro G, Iovanella A (2015) Organizing collaboration in inter-organizational innovation networks, from orchestration to choreography. Int J Eng Bus Manag 7(24)

Ferraro G, Iovanella A (2016) Revealing correlations between structure innovation attitude in inter-organisational innovation networks. Int J Computat Econ Econometrics 6(1):93–113

Ferraro G, Iovanella A, Pratesi G (2016) On the influence of nodes' characteristic in inter-organizational innovation networks structure. Int J Computat Econ Econometrics 6(3):239–257

Gloor P (2006) Swarm creativity, competitive advantage through collaborative innovation networks. Oxford University Press, Oxford

Hu T, Andrew AS, Karagas MR, Jason H, Moore JH (2015) Functional dyadicity and heterophilicity of gene-gene interactions in statistical epistasis networks. BioData Mining 8:43. doi:10.1186/s13040-015-0062-4

Park J, Barabási A-L (2007) Distribution of node characteristics in complex networks. Proc Natl Acad Sci U S A 14(46):17916–17920

Wang XF, Chen G (2003) Complex networks: small-world, scale-free and beyond. IEEE Circuit Syst Mag 3(1):6–20

White DR, Harary F (2001) The cohesiveness of blocks in social networks: node connectivity and conditional density. Sociol Methodol 31(1):305–359

Part III
Patterns

Chapter 9
Workshop Generator Patterns: A Supporting Tool for Creating New Values in a Workshop

Yuma Akado, Masafumi Nagai, Taichi Isaku, and Takashi Iba

9.1 Introduction

When the discussion on how to make workshops effective is being made, the existence of facilitator and its role is often the main topic. The function of a facilitator is to make sure the process work smoothly and support participants while being neutral (Gottesdiener 2003). Having a skilled facilitator to facilitate the workshop is said to make the workshop successful and meaningful.

On the other hand, a leader who leads the group is necessary when making a innovative team. Small group researches have been made over the decades, and as well as its process, the role of leader has been discussed as well (Levine and Moreland 1990).

There are also discussions on how collaboration can create innovative ideas. One states that collaborating could make a group more creative than working individually, under certain settings (Sawyer 2008). The research on collaborative innovation network, also known as COINs, shows that a network of people online can become innovative as well, when the team is working toward a shared goal (Gloor and Cooper 2007).

What we call a "generator" contains the characteristics of each role shown above. A generator's most important goal is to create new values while involving others into the process. It is somewhat close to facilitator, but a generator does not just support and rely on participants. A generator can be a leader in a way, since the generator also acts as a participant and gets involved into the discussion.

Y. Akado (✉) • T. Iba
Faculty of Policy Management, Keio University, Tokyo, Japan
e-mail: s13015ya@sfc.keio.ac.jp

M. Nagai • T. Isaku
Graduate School of Media and Governance, Keio University, Tokyo, Japan

© Springer International Publishing Switzerland 2016
M.P. Zylka et al. (eds.), *Designing Networks for Innovation and Improvisation*,
Springer Proceedings in Complexity, DOI 10.1007/978-3-319-42697-6_9

However, a generator thinks collaboratively to enhance the collaboration of the team while working together, not just leading the way to reach the desired goal.

In this paper, we will focus on the role of "workshop generator," who leads the group through the inquiry process to create new values in a workshop. The term "workshop" is used in this paper for short-term, around 2–4 h gathering, which contains brainstorming and discussions. The paper will describe the role of workshop generators in detail with examples from our practices and the *Workshop Generator Patterns*, and then show how *Workshop Generator Patterns* can be used and how it affects on the behavior of workshop generators.

9.2   The Workshop Generator

Over recent years, Iba laboratory has been holding two types of workshops to come up with ideas, at times for solving problems. In both workshops, a workshop generator works together with participants per work group, and aims to create new values by collaborating together (see Fig. 9.1).

Future Language (Iba 2015) workshops are held together with UDS Ltd. to gather ideas from participants for the space or the architecture UDS Ltd. is going to design (Honda et al. 2015). In the workshop, participants talk about their ideal space or architecture and also talk about the problems that they currently have. After that, the participants think about how they can realize the ideal vision and name the solution, which becomes Future Words.

While pattern mining workshop has been a place where participants write patterns within the short amount of time (Rising 1999), "Generative Beauty Workshop" is another type of pattern mining workshop, focusing on collecting knowledge as hints for patterns (Akado et al. 2015). In the workshop, participants share their own tips as well as concerns. If there are no existing solutions for the concerns brought up, participants seek for it through discussion.

Even though these workshops aim to create new and innovative ideas, some workshops have ended with results that were predictable. The ideas are the result which comes out from this complicated mixture of the types of participants we had

Fig. 9.1 Generators in workshops

that day, the number of people, the theme we worked on, and so on. While it is true that these aspects can affect the result, since workshop generators are in each work group during the workshop to lead and create the situation, supporting workshop generators could be a way to improve the workshop and end up with better results.

That is what motivated us to create the *Workshop Generator Patterns* as a supporting tool for workshop generators. From the 40 patterns of *Generator Patterns* that talks about generators in a broad sense, we picked out 9 patterns that are most usable for workshop generators.

9.3   The Workshop Generator Patterns

We have created the *Generator Patterns* through collaborative introspection and holding mining interviews (Iba and Yoder 2014). The interviewees included not only the people who have experienced workshop generators, but also teachers who interact with students and conducted classes as inquiry projects, or a person who have organized a cooking session where people cook and interact with each other. We wrote them down as patterns, and created a network-like structure between them as a pattern language.

The *Workshop Generator Patterns* are 9 patterns picked out of 40 *Generator Patterns*, which are most related to workshop generators who lead the work group during a workshop. Out of nine, three are abstracted patterns and each has two patterns that are more into detail (Fig. 9.2).

Patterns describe things that a workshop generator should keep in mind, or actions he/she should take (Table 9.1). For example. the abstracted patterns talk about how they should set the goal to create *A New Perspective (1)* through the workshop, not forget to *Put yourself Inside (4)*, but with a *Creating Together (7)* mindset.

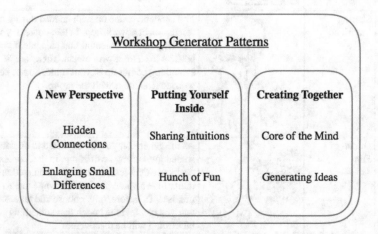

Fig. 9.2 Overall view of the workshop generator patterns

Table 9.1 Thumbnail of the workshop generator patterns

No.	Pattern name	Pattern illustration	Context, problem, and solution
1	**A New Perspective**		You are involved in an inquiry process as the leader of inquiry. In that context, since the goal of the group is not clear, there will be not much left to remember afterward even if interesting discussions occur. Therefore, create a new perspective through the inquiry that would change the way the participants would see and think about the world.
2	Hidden Connections		You are discussing an idea or comparing ideas with the participants to create *A New Perspective (1)*. In that context, you would not be able to provide a new discovery and end with an obvious result if you organize the ideas by existing categories or let ideas flow by deciding it is just not related to one another. Therefore, find hidden connections between elements that seem unrelated by highlighting similarities and thinking analogically.
3	Enlarging Small Differences		There are words that repeatedly used in the discussion to create *A New Perspective (1)*. In that context, the discussion will end with an impossible result if the word's definition is kept unclear. Therefore, notice abstract words repeatedly used by participants, and find small differences in its meaning between the participants.
4	**Putting Yourself Inside**		You are going to participate as a generator. In that context, being a facilitator keeps you out of the table and will make it difficult to find the meaning to participate in that certain activity. Therefore, get actively involved with the inquiry with your own curiosity and creativity activated.
5	Sharing Intuitions		You suddenly came up with an idea during the discussion at a workshop. In that context, you will let go of the potential that idea has if you hesitate to share it with others. Therefore, if something comes up in your mind, do not be worried to share it with the group.
6	Hunch of Fun		As the generator, you are expected to decide something during a workshop. In that context, the decision might lead to an easily predicted result rather than a value that could change one's mindset. Therefore, only choose and take actions that you can feel the hunch that something interesting will be discovered.

(continued)

Table 9.1 (continued)

No.	Pattern name	Pattern illustration	Context, problem, and solution
7	**Creating Together**		You are trying to weave an inquiry story with the participants in order to create *A New Perspective (1)*. In that context, the group would not be able to think out of the box if each participant or the generator holds on to the knowledge they already have. Therefore, engage in a zero-base creation of a new concept with the participants to provoke group creativity.
8	Core of the Mind		You are pulling out thoughts from participants in order to bring out characters and make them feel at ease. In that context, it is difficult to reflect each comment when participants are not fully expressing their feelings. Therefore, if you feel someone is not being able to express their thoughts accurately, listen carefully, ask questions, and provide keywords to help them say what they really mean to.
9	Generating Ideas		You want to assist participants to release their creativity through the dialogue. In that context, unique ideas from participants are not much created. Therefore, actively interact with the participants to help them generate ideas.

9.4 The Usage of Workshop Generator Patterns

The purpose of creating the *Workshop Generator Patterns* is to share the knowledge with others and improve the workshop generating skill. In order to see how the patterns affect workshop generators, we provided the patterns for the workshop generators who participated in two Future Language workshops in a row in February 2015. This section shows what we have found out from the received feedback.

9.4.1 Designing the Action During Workshop

Through the feedbacks, we have found out that patterns have affected workshop generators when designing their actions to take during the workshop. Below is a passage from the feedback written by one of the workshop generator:

> …*I kept in mind the* **A New Perspective** *pattern and tried to explain to the participants the purpose of the workshop, which was to come up with ideas to create a dormitory where students can enjoy their student life experience as a whole. I think this worked well, for group came up with not only ideas about the functional aspects of the building, but also about activities and interaction between students.*

This shows that looking through the *Workshop Generator Patterns* beforehand helps workshop generators to remember the patterns during the workshop, and therefore help them design their action. The patterns are useful when reviewed right before the workshop, with or without a discussion among the workshop generators using the patterns.

9.4.2   Looking Back and Planning the Actions by Patterns

We asked the generators to look back how they did as a generator. Below is a passage written by one of the workshop generators:

> *...One of the examples in which I felt like an effective generator, is when the overall atmosphere of our group was very intimate and warm by **Putting Yourself Inside**. With the help of some snacks, my outgoing and amiable characteristics have helped everyone to create a casual group climate. Particularly, my small talks and jokes related to my international background seemed to have inspired a sense of companionship. Another example is when I was able to **Generating Ideas**. Although the members did not seem to have a clear idea of their ideal student life in the beginning of the workshop, my questions inspired them to bring out ideas based on their dorm life. ...*

By using patterns as vocabularies, workshop generators were able to look back the effective action they took during the workshop. As well as looking back, they can also plan out their following session by using patterns as hints:

> *...As Future Language aims to inspire students to visualize and articulate their visions for the future, my goal for the next workshop would be **A New Perspective**. I would like to keep a forward-thinking mindset and help participants to reframe existing ideas and create new ones. During the workshop, I hope to value **Creating Together**. While it is easy that a generator could become a facilitator and lead the entire learning process, I would like to remind myself to be aware of non-verbal cues that may represent that the participants want to express something. ...*

The result shows that patterns work as inspiring vocabularies that can be used with normal words. Through looking back and creating plans by patterns, workshop generators are able to understand patterns more deeply than just reading them, and the patterns are kept in their mind as well.

9.4.3   Feedbacks on Patterns

As the final question for the feedback, we asked the workshop generators for comments on using the patterns. We received answers such as:

- Reading the *Workshop Generator Patterns* helped me reflect on the first workshop, and think about the things I need to improve on.
- It helped me make action plans for the next workshop.
- There are patterns that should be shared later on rather than reading it beforehand, as those patterns might limit the generator's thoughts.

- Due to the abstractness of the solution, I had a hard time grasping it and applying it in the workshop.
- Even though I was not able to read the patterns during the workshop, reading and understanding the main points to keep in mind beforehand helped me adjust the way of leading the discussion.

These comments show that the nine patterns were helpful in a way to look back how one did in a workshop and to plan ahead. On the other hand, some comments pointed out that few patterns were too abstracted in order to put them into action. Adding more concrete patterns into the structure could perhaps cover this problem.

In addition, some pointed out potential patterns that they felt in need. For example, one pointed out the need of a pattern about deepening questions, and another suggested a pattern on how to maintain participation throughout the entirety of the workshop.

9.5   Conclusion

In this paper, we have proposed *Workshop Generator Patterns* as a supporting tool for workshop generators. Through actually using the patterns, we have found that it is useful, but at the same time need some improvement. From the works we asked the generators to do, it has been shown that patterns work as vocabularies, useful in designing action during the workshop, and looking back as well as planning the next action. The comments we received tell how it was useful, but there are sequences that are not shown in the structure and also that there are potential patterns as well.

The result tells us that the *Workshop Generator Patterns* is not yet complete, and that it already has some potential patterns that are unwritten. Through further research, we aim to improve the *Workshop Generator Patterns* by restructuring the patterns by adding more patterns, and come up with different ways to make the patterns more effective as a supporting tool for workshop generators.

Acknowledgements   We firstly thank Jei-hee Hong for creating the patterns together with us. In addition, we thank the workshop generators who kindly replied to our research questionnaire: Sumire Nakamura, Eri Shimomukai, Ayaka Yoshikawa, Yuji Harashima, Shiori Shibata, Tomoki Kaneko, and Tsuyoshi Ishida.

References

Akado Y, Kogure S, Sasabe A, Hong J, Saruwatari K, Iba T (2015) Five patterns for designing pattern mining workshops. In: Proceedings of the 20th European conference on pattern languages of programs, Kaufbeuren, Germany, July 2015
Gloor PA, Cooper S (2007) Coolhunting: chasing down the next big thing. AMACOM Div American Mgmt Assn, New York

Gottesdiener E (2003) Requirements by collaboration: getting it right the first time. IEEE Softw 20(2):52–55

Honda T, Nakagawa K, Iba T (2015) Collaborative design of workplace with future language. Paper presented at the international conference on collaborative innovation networks 2015, Tokyo, Japan, March 2015

Iba T (2015) Future language as a collaborative design method. Paper presented at the international conference on collaborative innovation networks 2015, Tokyo, Japan, March 2015

Iba T, Yoder J (2014) Mining interview patterns: patterns for effectively obtaining seeds of patterns. Paper presented at the 10th Latin American conference on pattern languages of programs, São Paulo, Brazil, November 2014

Levine JM, Moreland RL (1990) Progress in small group research. Annu Rev Psychol 41(1):585–634

Rising L (1999) Patterns mining. In: Handbook of object technology, CRC Press, Boca Raton

Sawyer K (2008) Group genius: the creative power of collaboration. Basic Books, New York

Chapter 10
Design Patterns for Creative Education Programs

Norihiko Kimura, Hitomi Shimizu, Iroha Ogo, Shuichiro Ando, and Takashi Iba

10.1 Introduction

In recent education, an education style that cultivates learners' creativity is required; this is a style of education in which learners construct their own knowledge through creating something and collaborating with others (Iba et al. 2012). LOGO programming (Papert 1980), tinkering (Resnick and Rosenbaum 2013), and FAB (Gershenfeld 2007) are all examples. Through such activities, the successive emergence of *discoveries* (the element of creation/creativity) occurs in the programs (Iba 2010). We call these programs "creative education programs". These programs or courses have been practiced in many fields; thus, more and more teachers are required to design such creative education in their own field.

But how should they go about designing creative education? According to Christopher Alexander, who pursued the design process of architecture, the designer must trace his design problem to its functional origins and find some sort of pattern (form and context) in them (Alexander 1964); therefore, for designing creative education, it is necessary for teachers to understand the pattern that exists in education programs and design concepts.

In this context, we propose "Design Patterns for Creative Education Programs" based on the pattern language method to support the understanding of the education styles and its concepts.

N. Kimura (✉) • H. Shimizu • I. Ogo • S. Ando • T. Iba
Faculty of Policy Management, Keio University, Tokyo, Japan
e-mail: s13300nk@sfc.keio.ac.jp; iba@sfc.keio.ac.jp

© Springer International Publishing Switzerland 2016
M.P. Zylka et al. (eds.), *Designing Networks for Innovation and Improvisation*,
Springer Proceedings in Complexity, DOI 10.1007/978-3-319-42697-6_10

10.2 Classifying Education Patterns

Pattern languages are used to describe the design knowledge that exists in an area of profession. The pattern language method is used in many design fields; one of them is education. For instance, there are patterns for teaching computer programming (Pedagogical Patterns Editorial Board 2012), and we have also proposed some patterns for creative education (Shibuya et al. 2013; Harashima et al. 2014).

These patterns mainly deal with the behavior of teachers, but do not cover the design aspect of education programs well; moreover, the difference between designing behavior and designing programs has not been discussed explicitly. Here, we clarify this difference: according to the generations of pattern languages (Iba 2012) (Table 10.1), pattern languages that handle human actions are defined as Pattern Language 3.0; whereas, pattern languages that deal with non-material objects, such as education programs, correspond to Pattern Language 2.0. From this point of view, there are not enough education patterns that correspond to Pattern Language 2.0; hence, we aimed to create patterns that are more comprehensive by having patterns for designing education programs.

Therefore, in this paper, we propose a pattern language for designing creative education programs.

10.3 Design Patterns for Creative Education Programs

Here, we present "Design Patterns for Creative Education Programs". We first give a description of the process of creating the design patterns, and then show the summary of the 13 patterns.

10.3.1 Process of Creating Patterns

In this research we have used a two-stage process; mining interview and pattern writing, to create our education patterns (Fig. 10.1).

When creating a new pattern language, it is important to investigate the best experiences or best practices for collecting information to write patterns; we usually call this process "pattern mining". There are three general approaches to pattern

Table 10.1 Classify the education patterns

Pattern language	Object of design	Generation of pattern language
Design patterns for creative education programs	Education programs	Pattern Language 2.0
Pedagogical patterns for creative education patterns	Teachers' behavior	Pattern Language 3.0

Fig. 10.1 Process of creating patterns (*top*: mining interview, *bottom*: pattern writing)

mining (Kerth and Cunningham 1997). The first one is an introspective approach; where the author(s) investigate their own experiences. The second one is an artificial approach; where the investigator(s) study many related systems or architectures, notice the commonality and find the best practices from this analysis. The last one is a sociological approach; this approach studies experienced people who has the best practices. In this research, we used mining interview, one of the third approaches, because we wanted to mine patterns from teachers in different education fields from us and investigate how they design their courses.

We conducted mining interview based on "mining interview patterns" (Iba and Yoder 2014). These patterns recommend the interviewer(s) to ask questions along the following lines: solution, problem, and context (as we describe later, these three are key information for writing patterns). Firstly, the interviewer(s) ask experts the important practices or experiences that they want to share with colleagues. Then, they ask why those are important, in other words, what kind of problems will happen if they do not follow those practices. After that, they ask when or where those will be needed and when those problems will happen. Through such process, the interviewer is able to collect the information for pattern writing.

In our research at this point, we interviewed six experts, all professors of Keio University, in Japan (Table 10.2). All of them have years of experience teaching creative educational courses at Keio University (and other universities) in architecture design, computer programming, digital fabrication, etc. In the mining interview, we asked them what they really want to share about designing the educational programs, why that is important, and when or where that is needed. For mining patterns that could be used not only in Keio University, but also in other Japanese schools or overseas, we asked their experiences in teaching other universities (including overseas).

In the next stage, the pattern writing process, we wrote patterns in the same manner based on the information we collected in the mining interviews. All patterns have a *Context*, which shows the state we are in, a *Problem* we tend to run into in that state,

Table 10.2 Information about interviewees

Interviewee	Years of experience	Area of expertise
A	13	Design tool, design machine, personal fabrication, social fabrication
B	16	Fashion design critique and practice, design for social inclusion, interdisciplinary design research
C	16	Information design, experience design, human interface, mobile computing, urban computing, design aid system, creative activity support
D	7	Extreme, portable/light-weight architecture and energy design
E	11	Architectural design, algorithmic design, design process theory
F	13	Pattern languages, creativity, complex systems, social systems theory

a *Solution* that solves the problem, and *Consequence* that occurs when you practice the solution. Also, all patterns have a *Pattern Name* and a *Pattern Illustration*. By having a *Pattern Name*, a pattern can be a tool for thinking and communicating about the design (Iba 2011), and by having a *Pattern Illustration*, the readers of the pattern are more able to understand and remember the meaning of the pattern (Harasawa et al. 2014). Table 10.3 shows an example of the patterns, "*Improvement by Esquisse Check*".

10.3.2   The Overall View of Patterns

Table 10.4 shows the summary of "Design Patterns for Creative Education Programs". At this stage, there are 13 patterns. If the programs have these patterns, the successive emergence of discoveries occurs easily; therefore, the creation of learners' will be active (Iba 2010; Iba et al. 2012).

10.4   Show the Education Programs with Its Design Concepts

As we have already seen in the previous section, these design patterns deal with education programs, not the teachers' behavior. Therefore, people who read those patterns can use them for understanding the design concept of courses and programs, and reuse them in their context. In this context, we propose the way to show the syllabuses with its patterns. As an example, we looked at "Basic Design Studio", a course held in Keio University. We also propose the idea that using patterns on online open education tools will increase the teachers' opportunity to design their own original courses.

Table 10.3 A detailed example of "*Improvement by Esquisse Check*"

Pattern number	No. 4
Pattern name	**Improvement by Esquisse Check**
Pattern illustration	
Context	As the course goes on, the quality of students' outputs and ideas will vary greatly
	▼In this context
Problem	**If we ignore the situation, the gap between students gets bigger and bigger.** Some students move on smoothly, while others may be struggling. Adjusting to struggling students is not ideal since it will slow down the progression of the course. Although it would be best to look after every student during the class, it may be difficult to do so due to the number of students and limited time
	▼Therefore
Solution	**Check each student's Esquisse outside of class.** An esquisse check is the standard method of architecture design courses for students to make a plan on their project with the teacher 1-on-1. Make the struggling students come to the teacher outside class. Check the present problem on the spot, and lay out the idea together. After the check, the student will approach the project with the policy they planned with the teacher
	▼Consequently
Consequence	We can push up students who were falling behind. The ideas, projects that made a rapid progress by the check, become the center of attention in the next course. However, this method forces teachers to spare much time, so they need to be prepared for that. If the teacher has a reliable Student Assistant or Teacher Assistant, the teacher may depend on them.

10.4.1   Show the Programs with Its Patterns

By having design patterns, it is possible to clarify the design concept of the programs. Normally, the syllabuses show how the courses are programmed; however, these syllabuses do not show the underlying design concepts. For clarifying the design concepts, it is necessary to disassemble the program to each element, and show those design concepts; therefore, we propose showing the syllabuses with the design patterns, such as in previous section; thereby enabling us to understand the elements of the program and the underlying design concepts at the same time. As a result, less experienced teachers can use the patterns to refer to experts' course design and make their course better. As for students, since the design concept of the course is clarified by patterns, they also are able to make a learning plan of the course.

Here we take one course as an example. There is a course called "Basic Design Studio" in Keio University. This is a basic architecture design course for students in lower grades to learn architecture design. Despite it being one of the hardest courses

Table 10.4 Summary of patterns

No.	Pattern name	Pattern illustration	Summary
1	Abstract the essence		Students research and select two existing works as father and mother. Then, students read out the essence of them (like DNA), and create new ideas that inherit those essences
2	Vote for the attractive		There is not enough time for every student to give a presentation. Therefore, students and teachers take a vote on the attractive work they want to hear, and decide the order of it by the polling score
3	Visible efforts		Make everyone's project submission status visible to all students to allow them to see how much other students are working on the project to realize the gap between others
4	Improvement by Esquisse Check		If the teacher does not have enough time during class to look at every student's work and ideas, take time outside class to discuss their problems individually
5	Free time union		In the course that requires group work, form the group based on their free time for all group members to meet and work on the project together
6	Kick-off party		For good collaboration, the start of the group is very important. Therefore, the first task for the group is to go somewhere, like a restaurant, to hold a kick-off-meeting
7	Group's snapshot		When teachers want to see how every group is working, assign each group an activity report with a picture that shows what and how the group is working
8	Polyphonic discussion		If only certain students speak up during class, and you want every voice to be heard, make a small group and talk about the matter, then, share the group's discussions to the whole class

(continued)

Table 10.4   (continued)

No.	Pattern name	Pattern illustration	Summary
9	Free style presentation		To encourage students to be open-minded, make no limitations for the presentation of the project, or the format of it, so students can output freely
10	Frontier dialogue		For students to see the frontiers beyond the course, invite a special guest in the frontier and hold a dialog between the guest and the teacher in the class
11	Special stage		Make it a grand final presentation for the students so that it reflects the good work done and conclude the hardships of their project
12	Critiques for teachers		To get feedback so that teachers can further develop the class, invite other teachers (same school or other school), or specialists to the final presentation as guest critiques
13	Ending party		After the final presentation, hold a party to celebrate the successful completion of the course with both teachers and students

in the campus, it is very popular among students for mastering many skills. The course is programmed as below: by disassembling the program to each element (patterns), it is possible to understand what kind of knowledge is used to build the course (Fig. 10.2). By understanding the intensions of the design, it enables us to reuse the elements of the popular "Basic Design Studio" class.

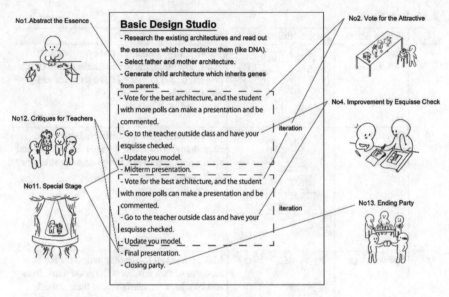

Fig. 10.2 The program of "Basic Design Studio" with its patterns

10.4.2 Show the Open Education Contents with Its Patterns

Recently, open education is being promoted by online tools such as Open Course Ware, and MOOCs (Massive Open Online Courses): good example is Open Educational Resources.[1] Since these online tools open the education contents, our opportunities to learn increased; however, it cannot be said that those online tools open everyone to design own education programs; this is because what is open online are the results of design, syllabuses and lecture notes, for instance, and these are distributed without its design concepts.

In this context, we propose the idea that those online tools open the courses with its patterns, so that the design concepts will be opened at the same time; therefore, it enable teachers to reuse the patterns and design original education programs; hence the opportunities to design original courses will be increased. As a result of opening the (creative) education programs with its patterns, open education cultivates not only students' creativity, but also teachers' creativity to design creative education.

10.5 Conclusion

In this paper, we proposed "Design Patterns for Creative Education Programs". These patterns contain the form of creative education programs and its design concepts, and told us how the programs should be designed. Therefore, people who

---

[1] https://www.oercommons.org

read the patterns can understand the good design of courses, and reuse the ideas of patterns for designing their own courses.

However, our patterns are not complete. In the current stage, our interviews covered a limited range of education fields, architecture design, fashion design, computer programming, etc. For the purpose of more comprehensive patterns that could help designing creative courses in many fields, we need to do mining interview for more experts in many fields, and find more various patterns. Furthermore, we also need to clarify the relation between other learning and education patterns, and study the usage of all patterns combined together.

Acknowledgments We thank all the interviewees who kindly responded to our request: Shohei Matsukawa, Daijiro Mizuno, Yasuto Nakanishi, Takuya Onishi, and Hiroya Tanaka. We also appreciate the support of Ayaka Yoshikawa who gave us much advice and translated our paper into English.

References

Alexander C (1964) Notes on the synthesis of form. Harvard University Press, Cambridge

Gershenfeld N (2007) Fab: the coming revolution on your desktop—from personal computers to personal fabrication. Basic Books, New York

Harasawa K, Miyazaki N, Sakuraba R, Iba T (2014) The nature of pattern illustrating: the theory and the process of pattern illustrating. In: Proceedings of the 21st conference on pattern languages of programs (PLoP'14), The Hillside Group, USA, October 2014

Harashima Y, Kubota T, Iba T (2014) Creative education patterns: designing for learning by creation. In: Proceedings of the 19th European conference on pattern languages of programs (EuroPLoP'14), ACM, New York, USA, July 2014

Iba T (2010) An autopoietic systems theory for creativity. In: Procedia—social and behavioral sciences, vol 2(4), pp 6610–6625

Iba T (2011) Pattern Language 3.0 methodological advances in sharing design knowledge. Paper presented at COINs2011: the international conference on collaborative innovation networks 2011, Basel, Switzerland, Sept 2011

Iba T (2012) Pattern Language 3.0: writing pattern languages for human actions. In: Invited talk in the 19th conference on pattern languages of programs, 2012

Iba T, Yoder J (2014) Mining interview patterns: patterns for effectively obtaining seeds of patterns. In: Proceedings of the 10th Latin American conference on pattern languages of programs (SugarLoaf PLoP'14), São Paulo, Brazil, Nov 2014

Iba T, Ichikawa C, Sakamoto M, Yamazaki T (2012) Pedagogical patterns for creative learning. In: Proceedings of the 18th conference on pattern languages of programs, October, 2011

Kerth NL, Cunningham W (1997) Using patterns to improve our architectural vision. IEEE Softw 14(1), Seite 53–59

Papert S (1980) Mindstorms: children, computers, and powerful ideas. Basic Books, New York

Pedagogical Patterns Editorial Board (2012) Pedagogical patterns: advice for educators. Joseph Bergin Software Tools, San Bernardino

Resnick M, Rosenbaum E (2013) Designing for tinkerability. In: Honey M, Kanter D (eds) Design, make, play: growing the next generation of STEM innovators. Routledge, New York, pp 163–181

Shibuya T, Seshimo S, Harashima Y, Kubota T, Iba T (2013) Educational patterns for generative participants: designing for creative learning. In: Proceedings of the 20th conference on pattern languages of program (PLoP'13), The Hillside Group, USA, Oct 2013

Chapter 11
Pattern Objects: Making Patterns Visible in Daily Life

Takashi Iba, Ayaka Yoshikawa, Tomoki Kaneko, Norihiko Kimura, and Tetsuro Kubota

11.1 Introduction

Pattern language is one of the methods to promote creativity in individuals and collaborative activities. It is a way of problem solving, and also functions as a common vocabulary that has been shared through various forms such as books, papers, cards, and Web pages. However, little has been discussed whether these forms of sharing are enough to actually encourage people to put the patterns to use in their daily lives. In this paper, we first provide an overview of the pattern language method and conventional media for sharing pattern languages. Then, we propose the concept of *pattern objects* as a way to make patterns more visible in daily life environments, and demonstrate five prototypes.

11.2 Pattern Languages and Their Media for Sharing

Pattern languages are used to describe experts' knowledge in ways that can be shared and used by others. A pattern describes expertise in problem solving for a particular context, and has a pattern name so that this knowledge can be communicated. Patterns also have some visual aids, such as pictures, diagrams, and illustrations.

T. Iba (✉) • N. Kimura
Faculty of Policy Management, Keio University, Tokyo, Japan
e-mail: iba@sfc.keio.ac.jp; s13300nk@sfc.keio.ac.jp

A. Yoshikawa • T. Kaneko
Faculty of Environment and Information Studies, Keio University, Tokyo, Japan

T. Kubota
Keio Research Institute at SFC, 5322 Endo, Fujisawa 252-0882, Japan

© Springer International Publishing Switzerland 2016
M.P. Zylka et al. (eds.), *Designing Networks for Innovation and Improvisation*,
Springer Proceedings in Complexity, DOI 10.1007/978-3-319-42697-6_11

The pattern language method is used in many fields. The original idea of pattern language was proposed by Christopher Alexander, an architect who along with his colleagues wrote a book that contained 253 patterns for architecture design (Alexander et al. 1977). This method was then adopted in the field of software design (Beck and Cunningham 1987). The "*Design Patterns: Elements of Reusable Object-Oriented Software*" (Gamma et al. 1995) is a pattern book that is especially well known in the software field. Since then, this method has been applied to describe human action, and many pattern languages have been created and published (Coplien and Harrison 2004; Manns and Rising 2004; Pedagogical Patterns Editorial Board 2012). We also created and published some pattern languages about human action on topics like learning (Iba and Iba Laboratory 2014a), presentation (Iba and Iba Laboratory 2014b), collaboration (Iba and Iba Laboratory 2014c), social innovation (Shimomukai et al. 2015), disaster prevention (Furukawazono and Iba 2015), and living well with dementia (Iba and Okada 2015).

Pattern languages, in any topics, have generally been presented as reading materials. Many patterns have been published in the form of books. In the software community, conferences of pattern language (PLoP and other PLoPs held all over the world), have taken place over the past 20 years, where patterns have been shared as academic papers. Patterns have also been shared through the Web and made available for anyone to edit, such as Ward Cunningham's Portland Pattern Repository,[1] which contained his software design patterns in the first ever wiki, or reader-modifiable Web pages.

In recent years, patterns have also taken a form of pattern cards. Pattern cards, such as Iba Lab's Learning Patterns Cards and Oregon's Group Works[2] are new tools for using pattern languages in collaborative settings such as dialog workshops (Iba 2014). There are also the Fearless Journey Game cards, which expresses patterns in the form of a game.

Recording and sharing patterns through printed and online materials are practical ways for users to access and read them. However in many cases, the situation in which pattern users read the pattern versus where they actually exercise the pattern differs. In fact, aside from a few software design patterns, you do not read and practice the patterns simultaneously.

It is said that Christopher Alexander, who developed the method of pattern language in the field of architecture told his students to have his patterns memorized in their heads before coming to the construction site. Although we do not know whether this was actually said, we can see from this story that patterns must be remembered in some way before they can be put into practice.

Patterns must also be recalled at the right timing in order for them to be functional. Pattern usage is initiated by the identification of the context; a person identifies a particular situation that he or she is in, recalls some patterns in their

---

[1] http://www.c2.com/cgi/wiki

[2] http://groupworksdeck.org/about

mind that have similar contexts, and then selects the pattern that best suits their situation. If the person fails to recall the suitable pattern for a given situation, the option of using a pattern to prevent or solve a problem becomes invalid. Not to mention, the very first step of identifying the situation requires meta-recognition, which is not always a simple task. Failing to recall the right pattern means that patterns would not be used, thus making it likely for a problem to occur.

So, what is a better way to share patterns? Our answer is to make objects that express patterns, which can be placed around our living environments to help us recall the desired pattern for a certain situation. These are what we call *pattern objects*.

11.3   Pattern Objects

Pattern objects are objects that express patterns to help us recall the right pattern for the right situation. *Pattern objects* could be tools and goods that we use in our everyday lives that are modified to represent a pattern, or completely newly invented objects. Pattern users should first read a written pattern, and then place a *pattern object* that embodies the pattern at a place where the desired action may take place. The *pattern object* will then function as a trigger for people to recall and use the suitable pattern when they need it.

Pattern objects should make patterns more effective in our daily lives. For instance, if you have a cooking pattern about how to heat something up, you can place a *pattern object* that expresses the pattern somewhere near the stovetop so that you can keep it in your mind as you cook. Or, if there is a pattern about how to cut ingredients, it may be helpful to have that pattern engraved on the corner of your cutting board. The examples above are about cooking, but *pattern objects* of any other field can exist in different shapes and forms that are suitable to each environment. It is important to mention that *pattern objects* must be designed in a way that makes the pattern usable in a particular environment. They are not merely objects printed with a pattern name or illustration as "logos". Each pattern language has its own settings that fit with their theme, and each individual pattern has its own functional object that expresses the pattern.

This idea of putting patterns around our environment (instead of relying solely on memory) is logical from a cognitive science viewpoint. In the theories of cognitive science, humans are said to be able to execute things better if bits of their memory are engraved in their surroundings as reminders (Lave 1988). Similarly, patterns are something to be read beforehand as base knowledge, and *pattern objects* are the forces from the outside that remind and encourage people to put the patterns into practice.

What we propose in this paper is new to the pattern language field, but is a rather common practice in our everyday lives. For instance, writing down reminders on sticky notes and putting it on our desks is a similar action that we practice regularly.

In the workplace, we see examples of companies like IDEO posting their brainstorming policy on their wall. Schools like MIT also have statues of their founder, as reminders of their mission and principle.

When looking at such examples, it is quite a mystery that pattern languages have only been shared through the medium of writing. Our goal is to now express the patterns through three-dimensional objects that can be placed around our living environments. The following section introduces some prototypes of *pattern objects* we have created thus far.

11.4 Prototypes of Pattern Objects

We present five prototypes of *pattern objects*: cutting board, paper clips, snack box, refrigerator magnet, and survival basket, each expressing a pattern (or patterns) that is suitable to the context. The objects presented here are mere prototypes, but they give an idea of how *pattern objects* can be used to make pattern languages more effective in the future.

11.4.1 Cutting Board

The "Uniform Bites" Cutting Board embodies the "*Uniform Bites*" pattern from the *Cooking Patterns* (Akado et al. 2016), which says, "You are cooking with ingredients of varying sizes. In this context, a dish that uses ingredients with varying sizes tends to lack uniformity. Therefore, create a uniform texture by cutting your ingredients in the same size."

Since the pattern deals with how to cut ingredients, we engraved the pattern onto a cutting board. This allows the cook to see the pattern and be reminded when he or she is preparing the ingredients. Figure 11.1 shows this *pattern object* and how it is used in daily cooking.

Fig. 11.1 A cutting board for creating uniform texture in a dish

Fig. 11.2 Paper clips, for reminding the important patterns when finishing up a writing process

11.4.2 Paper Clips

We also made Pattern Clips, which act as reminders of patterns through a writing process. For our first prototypes, we picked out several patterns that focus on polishing and finishing a piece of writing from the *Learning Patterns*, the *Presentation Patterns*, and the *Collaboration Patterns* (Fig. 11.2).

For example, "*The First-Draft-Halfway-Point*" pattern from the *Learning Patterns* (Iba and Iba Laboratory 2014a) says, "You are writing your ideas to share them with others. In this context, the initial draft is not suitable to be read by others. Therefore, after finishing an initial draft, improve it objectively, considering whether readers will easily understand."

The "*Exploration of Words*" pattern from the *Presentation Patterns* (Iba and Iba Laboratory 2014b) says, "The *Storytelling* that conveys your *Main Message* is decided, and now you are making your presentation. In this context, to avoid using dull or overused expressions you are tempted to use unfamiliar words, which may be difficult for your audience to understand. Therefore, look for words and expressions that both you and your audience find attractive." These pattern clips function as reminders of important patterns that a writer should keep in mind.

11.4.3 Snack Box

The Pattern Snack Box is an example of a *pattern object* that embodies more than one pattern. This time, we created a wooden snack box with four different patterns engraved onto the sides (Fig. 11.3). One pattern we used is "*Brain Switch*" from the *Learning Patterns* (Iba and Iba Laboratory 2014a), which says, "You are creating an output, and you've made some progress. In this context, logical thinking is not sufficient to achieve a breakthrough without intuitive thinking and vice versa. Therefore, switch between the two modes of logical and intuitive thinking."

Fig. 11.3 Snack box that encourages creative thinking during collaborative work

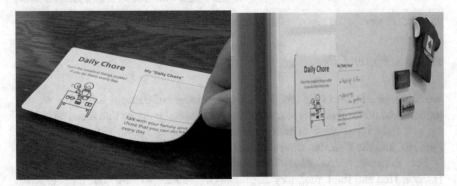

Fig. 11.4 Magnet on the refrigerator, reminding the daily chore for people with dementia

Another pattern is "*Ideas Taking Shape*" from the *Collaboration Patterns* (Iba and Iba Laboratory 2014c), which says "You have an idea you want to share with your teammates. In this context, the newer your idea is, the more people will not understand it. Therefore, visually shape your idea, so that others can see it while you explain it." This snack box serves as a refreshing reminder of important patterns during collaborative work.

11.4.4 Refrigerator Magnet

We also created a Daily Chore Magnet, which functions as a tool to help people with dementia practice the "*Daily Chore*" pattern from *Words for a Journey* (Iba and Okada 2015). This pattern says, "You increasingly need the help of others to do things for you. In this context, if you start to think you shouldn't do something on your own and should have everything done by others, you will start to become unable to do even the tasks that you can do now. Therefore, talk with your family and create a chore that you can do by yourself every day."

This magnet has space for users to write down their own daily chore, and could be placed on a refrigerator as a daily reminder for people with dementia (Fig. 11.4).

Fig. 11.5 Survival basket for regularly maintaining food supplies in preparation for natural disasters

11.4.5 Survival Basket

Finally, we created the Survival Basket using patterns from the *Survival Language* (Furukawazono and Iba 2015). On the bottom of the basket, we stitched the pattern "EXTRASTOCK" which is a pattern about stocking up food with longer expiration dates so that you have an emergency supply of foods and drinks in case of a natural disaster such as an earthquake. When the supply gets scarce, the pattern printed on the bottom will remind users to restock their emergency supply. We also put another pattern, "DAILY USE OF RESERVES" on the front side of the basket to encourage people to use the "EXTRASTOCK" of food supplies on a daily basis (Fig. 11.5). This will ensure that the products are not expired or forgotten by the time you actually need them. This basket may be placed around the kitchen area as a constant reminder to maintain emergency supplies, and will help people stay prepared for natural disasters.

11.5 Conclusions

In this paper, we proposed the idea of *pattern objects* and showed some prototypes. These *pattern objects* were designed to express patterns in familiar objects that we use in our daily lives. By placing these *pattern objects* in our workplaces and homes, we can be reminded of a certain pattern at the right place or time. However, these prototypes still need to be improved so that the objects fit well with the designated environment, and encourage people to actually take action. As one of the future works, we want to create *pattern objects* that express a pattern well enough to the point where users can understand it without reading a written pattern beforehand. Furthermore, we would like to test the effectiveness of pattern objects by using them in our daily settings and collaborative works.

References

Akado Y, Shibata S, Yoshikawa A, Sano A, Iba T (2016) Cooking patterns: a pattern language for everyday cooking. Paper presented at the 5th Asian conference on pattern languages of programs, National Taipei University of Technology, Taipei, 24–26 Feb 2016

Alexander C, Ishikawa S, Silverstein M, Jacobson M, Fiksdahl-King I, Angel S (1977) A pattern language: towns, buildings, construction. Oxford University Press, New York

Beck K, Cunningham W (1987) Using pattern languages for object-oriented programs. In: OOPSLA-87 workshop on the specification and design for object-oriented programming, Orlando, FL

Coplien JO, Harrison NB (2004) Organizational patterns of agile software development. Prentice-Hall, Inc., Upper Saddle River

Furukawazono T, Iba T (2015) Survival language: a pattern language for surviving earthquakes. CreativeShift Lab, Yokohama

Gamma E, Helm R, Johnson R, Vlissides J (1995) Design patterns: elements of reusable object-oriented software. Addison-Wesley Longman Publishing Co., Inc., Boston

Iba T (2014) Pattern language as media for creative dialogue: functional analysis of dialogue workshop. In: Baumgartner P, Sickinger R (eds) PURPLSOC: the workshop 2014, Krems, pp 212–231

Iba T, Iba Laboratory (2014a) Learning patterns: a pattern language for creative learning. CreativeShift Lab, Yokohama

Iba T, Iba Laboratory (2014b) Presentation patterns: a pattern language for creative presentations. CreativeShift Lab, Yokohama

Iba T, Iba Laboratory (2014c) Collaboration patterns: a pattern language for creative collaborations. CreativeShift Lab, Yokohama

Iba T, Okada M (eds) with Iba Laboratory, Dementia Friendly Japan Initiative (2015) Words for a journey: the art of being with dementia. CreativeShift Lab, Yokohama

Lave J (1988) Cognition in practice: mind, mathematics and culture in everyday life. Cambridge University Press, Cambridge

Manns ML, Rising L (2004) Fearless change: patterns for introducing new ideas. Addison-Wesley Professional, Boston

Pedagogical Patterns Editorial Board (2012) Pedagogical patterns: advice for educators. Joseph Bergin Software Tools, San Bernardino

Shimomukai E, Nakamura S, Iba T (2015) Change making patterns: a pattern language for fostering social entrepreneurship. CreativeShift Lab, Yokohama

Chapter 12
From Chefs to Kitchen Captains: A Leader Figure for Collaborative Networks in the Kitchen

Taichi Isaku and Takashi Iba

12.1 Introduction

This chapter will explore the possibility of collaborative networks in a rather unique environment: the kitchen. We will do this by introducing a new persona we have identified called the *Kitchen Captain*, who leads the collaboration in the kitchen.

Cooking, though widely conceived as either a household chore or a professional skill, is the simplest form of design that many of us go through on a daily basis. It is a creative process that requires abstract skills including generating ideas and solving problems. It may seem a highly complex set of skills, but many of us are able to overcome the difficulty with the process driven by hunger. What's better, these abstract skills, with a little help, are skills that can be applied and used in various situations outside the kitchen as well.

To further deepen the discussion, we will introduce a rather new concept called *CoCooking*. By introducing a collaborative aspect to cooking, its creative aspects are enhanced. By inviting multiple people into the kitchen—regardless of their ability to cook—the experience becomes a platform for co-creation and collaborative learning. The *Kitchen Captain* is merely the leader persona that appears in such a cooking.

The idea of the *Kitchen Captain* was derived from a superordinate persona called the *Generator*. The *Generator* is a rather new leader figure we have found to be present in collaborative networks who leads and iterates the process of a collaborative inquiry by involving the people around her into the process by enhancing their

T. Isaku (✉)
Graduate School of Media and Governance, Keio University, Tokyo, Japan
e-mail: tisaku@sfc.keio.ac.jp

T. Iba
Faculty of Policy Management, Keio University, Tokyo, Japan
e-mail: iba@sfc.keio.ac.jp

© Springer International Publishing Switzerland 2016   113
M.P. Zylka et al. (eds.), *Designing Networks for Innovation and Improvisation*,
Springer Proceedings in Complexity, DOI 10.1007/978-3-319-42697-6_12

creative desires. When the collaborative inquiry occurs in the kitchen through cooking, there the *Kitchen Captain* is present. The idea of the *Generator* along with her traits and skills are covered and formulated into a pattern language in another study by a team including the authors (Nagai et al. 2016). This chapter will go off on a tangent from this earlier study to consider the *Kitchen Captain* and inquires in the kitchen to propose a new form of cooking that focuses on enhancing team creativity and communication.

12.2 Background

As background information for our chapter, we will first go briefly over general trends on our relation with design and creation, and then we will get into more details by focusing our discussion on cooking.

12.2.1 From Industrialization to the Creative Society

For much of the past few centuries, we have lived in an age of consumption. Thanks to the invention of machines, we have enjoyed the luxury of consuming commodities at a relatively low price. However, more recently, some people have started to think such a lifestyle is not sustainable in the long run and are starting to question the fact we are eating the exact same foods and wearing the exact same clothing as our peers living on the opposite side of the planet.

Still, advances in technology see no limit, and more and more of our jobs are now being taken over by factories and computers who do a much better job than us. With such rapid changes occurring to our civilization, the way of living and the skills needed to survive are also changing. Many of us are now seeing value in a sustainable and a more personal style of living, where abstract and creative skills that machines and computers can't do for us become important. Iba capture this coming society as the *Creative Society*, where people create his or her own things, systems, or the way of understanding needed to live (Iba and Furukawazono 2013). This is in no means a nostalgic return to the past, but rather a new way of fulfilling life built upon the newest science and technology. In this emerging new society, we are in a need for a way to learn and teach these creative skills.

12.2.2 The Struggles of Cooking

The situation is nothing different in the context of cooking—especially in the home. Let alone the factory, cooking is often considered either a household chore or a service to pay for. Of course some people cook for pure joy, but this kind of cooking is limited to those who can do it as a personal hobby or a profession. The kitchen becomes a black box for many, and as much as we love taking out ready-to-eat dinners or defrosting frozen food, we start to close our eyes from who cooked the meal and how.

On the light side, with everything coming and going so fast, some of us are starting to question the way we eat. Instead of being beguiled by all the artificial flavors, we are now finding satisfaction in spending time to make our own food. This might sound like a return to the past, but it inherits many aspects from recent stages and builds upon their technology. After seeing some extremes, we are now trying to find the right balance.

At this recent phase where people are trying to cook again, we are facing new types of difficulties—namely, a lack of ideas. There is a huge gap between the population that can cook and the population that can't. There are some people who can open the refrigerator, get inspired from what's inside, and then cook a delicious dish from it. Then there are others who can't cook without following a recipe word-for-word. And, of course, there are the rest of us who need help just to step inside the kitchen. Our work with CoCooking is intended for all of these groups of people. Our goal is to define a fun, social, and creative form of cooking that puts people with a good mix of cooking skills together in the kitchen. Cooking—at least in the home—should be an improvised activity where collaborative learning and creative innovations occur on a daily basis.

12.3   CoCooking: A Collaborative Network in the Kitchen

CoCooking is a term we have coined to capture the collaborative networks that form inside a kitchen. It has the base word "Cooking" with the collaborative prefix "Co." The photos in Fig. 12.1 should give an image of the activity.

Though the idea may be simple, its possibilities see no limit. The analysis of its positive effects were covered in a different work, but to summarize, CoCooking has the following positive characteristics:

- *Acquisition of Cooking Techniques*: The mix between experts and novices in the same kitchen brings a good mix of skill and ideas, bringing novices a chance to learn cooking techniques while having fun and communicating with others.

Fig. 12.1 Pictures from past CoCooking sessions—the situation varies from home-cooking situations to educational and business use

- *Building Teamwork*: Cooking requires a lot of problem solving. When this process is done by a group, it naturally requires communicating and working together towards a common goal. Even if the participants don't have much in common, the existence of food creates a common topic to talk about, and the aroma of delicious food creates a warm atmosphere for conversations.
- *Nurturing Creativity*: Cooking is a creative process that requires abstract skills such as looking at available ingredients to generate ideas, making concepts, planning and coordinating steps, materializing the ideas, solving problems, and presenting the dish aesthetically. Cooking allows us to experience the sophisticated creative process that is otherwise hard to set up.

The best practice of CoCooking occurs when there is a good mix of skills among the participants, and the party does not exactly know what they are about to cook. With these properties, the collaborative networks tend to get concentrated, and innovation is likely to occur. Though held in a local and physical environment, these networks tend to show similar characteristics as a Collaborative Innovation Network (COIN) (Gloor 2006). The best way to become fond of the food we eat is to have experience cooking it ourselves. By experiencing the joy, effort, and the room for professionalism of cooking, not only can we support a healthy, creative, and sustainable lifestyle, we will also become able to appreciate food that are cooked by professional chefs.

In previous works, we have created tools that support a successful CoCooking session (Isaku and Iba 2015, 2016). These tools provided hints written in a pattern language format (Alexander et al. 1977), on how to make CoCooking more collaborative and creative. Over the process of this consecutive work, a leader-like persona existent in CoCooking sessions has come up. We named her the *Kitchen Captain*, the main focus of this chapter.

12.4 The Generator and the Kitchen Captain

The *Kitchen Captain* is the name given to a persona that we found to be existent when there is a successful CoCooking session going on. We can go on describing the characteristics of the *Kitchen Captain*, but doing so in an unsystematic way would be confusing. Instead, we will start by actually introducing a superordinate persona that encapsulates the *Kitchen Captain*—the *Generator*. We will first introduce this second persona, show our works with the idea, and then describe the *Kitchen Captain* from there.

12.4.1 The Generator as a Leader Figure in the Creative Society

The idea of the *Kitchen Captain* was created in the course of a larger project called the *Generator Patterns Project* (Nagai et al. 2016). The project's goal was to define a new kind of a leader persona called the *Generator*.

The *Generator* persona encapsulates the idea of a *Kitchen Captain*. Not limited to a kitchen setting, a *Generator* leads and iterates the process of a collaborative inquiry by involving the people around her into the process by enhancing their creative desires. This is not a role to replace any existing jobs, but rather a set of characteristics that a person in any position—for example, a chef—can show to make the surrounding atmosphere and people more creative.

In the *Creative Society*, we often must decide what to make—and what not to make—under certain situations. This ability of creating a best-fit solution under a context or an occurring problem is not achieved by choosing from ready-to-eat options, but rather requires a process perhaps similar to what Dewey defined as the process of inquiry (Dewey 1938). Dewy, along with Pierce and other pragmatists, highlighted that this process of inquiry goes beyond the system of thinking of an individual—forming a *Community of Inquiry*. The concept of "generator" comes from the idea that the role of a leader-like figure in this Community of Inquiry is to "generate" creativity within the members. The name also contains the metaphor that this person has an ongoing engine within herself that generates curiosity and ideas, just as an electric generator generates electricity.

12.4.2 The Generator Patterns

Pattern language is a method to scribe out the collective knowledge of a complex idea by breaking it down into a set of words—or *patterns*—that each describes one aspect of the idea. Then the set of patterns are organized in an understandable manner to define a coherent and emergent whole.

The *Generator Patterns* were created through in-depth interviews with seven people who we found to be showing leader characteristics that fit our image of a *Generator* which included the following: Chikara Ichikawa who is a teachers at an alternative school, Keibun Nakagawa who is the representative of the company UDS, which is a client-focused architecture planning and designing company, and Junichi Harajiri who works as a marketing designer.

The interview questions were centered on how they usually communicate with their team to generate ideas and enforce teamwork. We also asked each interviewee what kinds of things they would say, if they were to teach someone with the same profession to become better at the topic. The interview proceeded in a conversation style rather than in a one-question–one answer format. Though there was a set of questions we prepared, they were mere conversation starters, and we were free to go off on any tangent to deepen any topics that seemed interesting. This style of interviews is called a *Mining Interview* and is covered in depth in another paper by one of the authors (Iba and Yoder 2014).

In addition to the seven in-depth interviews, the five project members, including the authors, also gave our own experiences through introspective brainstorming and mutual interviews. The members included Masafumi Nagai with experience with coaching, Yuma Akado and Jei-Hee Hong with workshop facilitation skills (Akado et al. 2015), Takashi Iba with teaching experience university professor

0. Generator

1. Leader of the Inquiry	2. Putting yourself Inside	3. Releasing the Creativity
4. A New Perspective	**16. Original Style**	**28. Relaxed Atmosphere**
5. Hidden Connections	17. Professional Skills	29. Core of the Mind
6. Changing Filters	18. Original Tools	30. Faithful Guide
7. Enlarging Small Differences	19. Gift of Discoveries	31. Showing Earnest
8. Inquiry Tale	**20. Equal Participant**	**32. Generating Ideas**
9. Supple Plans	21. Thorough Discussions	33. Showing Interest
10. Posting Flags	22. Sharing Intuitions	34. Nurturing Ideas
11. Meaningful Coincidences	23. Life-Sized Person	35. Removing the Frame
12. Creating Together	**24. Creative Disposition**	**36. The Heartstrings of Curiosity**
13. Open Thoughts	25. Overlapping Curiosities	37. Vigorous Goal
14. Connecting Thoughts	26. Hunch of Fun	38. Leveraging Interests
15. Idea Billboard	27. Daily Inquiry	39. Afterward Plans

Fig. 12.2 The 40 patterns in the *Generator Patterns* shown as a list: three main categories with each category containing 12 other patterns that describe traits of a *Generator*

(Iba et al. 2012) and also as a workshop facilitator (Iba 2015), and Taichi Isaku with CoCooking facilitating skills (Isaku and Iba 2015, 2016).

Elements from these interviews and brainstorming were gathered and clustered through the KJ method, a bottom-up clustering method similar to the grounded theory (Kawakita 1967). Though the information is from a wide range of professions, this process allowed us to find the important essence common among the different types of leaders. The formed clusters were each written as patterns, and finally, the patterns were organized into a coherent whole.

If you look at Fig. 12.2, you can see that the 40 patterns in the pattern language are categorized into three groups. For this chapter, we will pick up six patterns—from the collection and introduce it specifically in the context of *Kitchen Captains*. The rest of the patterns can be read in Nagai et al.'s 2016 paper. Though the patterns are described in this chapter in a general context, they can all be applied to CoCooking situations in the kitchen.

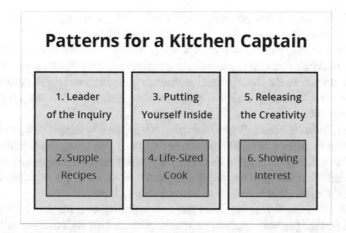

Fig. 12.3 The figure shows the relationship between the six patterns introduced in this chapter. Two patterns from three categories are shown

12.4.3 Patterns for the Kitchen Captain as a Kitchen Generator

When a chef—or someone with at least adequate cooking skills—shows traits as a *Generator*, she is called a *Kitchen Captain*. To explain this persona in a rather systematic way, we will introduce six patterns from the *Generator Patterns* in the context of CoCooking. The patterns will include one categorical pattern describing a rather abstract idea and one general pattern that provides a specific action that can be taken from each of the three categories in the language. These patterns are written mainly for informal occasions with cooking with family, friends, coworkers, etc. as the main target, but we think chefs in culinary kitchens can most certainly find some hints regarding the enhancement of creativity and building of teamwork. Figure 12.3 shows the semantic relationships between the patterns.

Patterns are written in a rather strict format with distinct sections including the *context* indicating when the pattern can be used, the *problem* which describes a difficulty that occurs in the context, the *solution* which describes an action that the reader can take in order to solve the problem, and the *consequence* or results that can be anticipated after applying the solution. However, due to spatial reasons, we will give a brief abstract that describes the essence of each of the patterns.

1. *Leader of the Inquiry*: When CoCooking with peers, you often wonder how you should interact with the other participants. Think of the opportunity as a new project and involve the people around you into the kitchen so you can lead a process of collaborative inquiry driven both by curiosity and hunger. This way, not only will the participants' hunger and curiosity be fulfilled, the group can build teamwork, become creative, and learn together.

2. *Supple Recipes*: When planning for a CoCooking, you might wonder what to cook: if you use a recipe, then the inquiry aspect of the cooking will be lost, but without a recipe, you are worried that things will get chaotic and uncontrollable. When building plans, you can prepare a recipe, but only use it as a guideline to set the direction. Be flexible and incorporate coincidences into the process during the actual cooking. This way you can prevent chaos and induce creativity at the same time.

3. *Putting Yourself Inside*: When becoming a Kitchen Captain, if you just become a normal facilitator or instructor, the necessity of you being the captain will be lost. Get actively involved with the inquiry with your own hunger and creativity activated. This way you can enjoy and get the most out of the experience, and it will induce the other members to become deeply involved also.

4. *Life-Sized Cook*: When you have the responsibility as a *Kitchen Captain*, you might feel the pressure that you have to be perfect. This isn't true—don't portray yourself as someone too special. A *Kitchen Captain* does not have to be a professional chef. If there is something you don't know, open up the discussion to the group. This way the group can become creative together.

5. *Releasing the Creativity*: When interacting with other people as the *Kitchen Captain*, you might not know how you should interact with the beginning cooks who have trouble coming up with ideas. Interact with them by inducing ideas from them, showing interest, and pushing their backs to join the actual cooking. This should nurture creative minds in the participants and will help build self-confidence in giving ideas and cooking.

6. *Showing Interest*: When CoCooking with a group with a good mix of skills, the beginning cooks might feel uncomfortable in the kitchen. Welcome ideas by everyone onto the counter by showing genuine interest and together as a group build on their idea. This way the group might end up with an unexpected but delicious dish, and it would also build the confidence of the novices.

12.5 Conclusion and Further Possibilities

This chapter explored a new leader persona in the kitchen called the *Kitchen Captain*. The kitchen, with cooking having so many creative aspects, can become a platform for cocreation and collective learning. In such a collaborative cooking—or CoCooking—the *Kitchen Captain* becomes an essential personality to optimize the experience. This chapter is part of a series of work with the mission of applying cooking as a way to enhance people's creativity. This work will continue with developing workshops, programs, and a suite of tools that potential and existing *Kitchen Captains* can use to enhance future CoCooking sessions.

References

Akado Y, Kogure S, Sasabe A, Hong J, Saruwatari K, Iba T (2015) Five patterns for designing pattern mining workshops. In: Proceedings of the collaborative innovation networks, Tokyo, eprint arXiv:1502.01142

Alexander C, Ishikawa S, Silverstein M (1977) A pattern language: towns, buildings, construction. Oxford Press, New York

Dewey J (1938) Logic: the theory of inquiry. Holt, Rinehart and Winston, New York

Gloor PA (2006) Swarm creativity: competitive advantage through collaborative innovation networks. Oxford University Press, New York

Iba T (2015) Future language as a collaborative design method. In: Proceedings of the collaborative innovation networks, Tokyo, eprint arXiv:1502.01142

Iba T, Furukawazono T (2013) Pattern language as media for the creative society. J Inf Process Manag 155(12):865–873

Iba T, Yoder J (2014) Mining interview patterns: patterns for effectively obtaining seeds of patterns. Paper presented at the 10th Latin American conference on pattern languages of programs, Sao Paulo, 9–12 Nov 2014

Iba T, Ichikawa C, Sakamoto M, Yamazaki T (2012) Pedagogical patterns for creative learning. In: Proceedings for the 19th conference on pattern languages of programs, ACM, New York

Isaku T, Iba T (2015) Creative cocooking patterns: a pattern language for creative collaborative cooking. In: Proceedings of the 20th European conference on pattern languages of programs, ACM, New York

Isaku T, Iba T (2016) Creative cocooking patterns: a pattern language for creative collaborative cooking part 2. Paper presented at the 21st European conference on pattern languages of programs, Irsee, 6–10 July 2016

Kawakita J (1967) Hassou Hou [the abduction method: for creativity development]. Chuo-Koron, Tokyo

Nagai M, Isaku T, Akado Y, Iba T (2016) Generator patterns: a pattern language for collaborative inquiry. Paper presented at the 21st European conference on pattern languages of programs, Irsee, 6–10 July 2016

Part IV
Social Media and Social Networks

Chapter 13
Measuring the Level of Global Awareness on Social Media

Peter A. Gloor, Andrea Fronzetti Colladon, Christine Z. Miller, and Romina Pellegrini

13.1   Introduction

What are the things that capture our attention? When walking down the street in our neighborhood, we might notice a "for sale" sign that was not there before or a newly planted flower bed. We tend not to notice things that have not changed, but if we see a broken window that was in one piece yesterday; it captures our attention. We wonder how it happened. Who might have done this and why? Most likely we will tell someone and ask if they saw it, too.

While this scenario unfolds at the neighborhood level, a similar phenomenon occurs at the global level. We use Twitter, Facebook, e-mail, and other online media to communicate what we "see" is happening in the world. We interact through multiple networks sharing information, opinions, and insights, in the process creating a collective awareness around the event (Sparrow et al. 2011). We participate in a process of collective sense making within a global community of people who share an interest in the things that we are interested in and whose lives are affected by the things that impact our lives.

Does an organization—and thus ultimately humanity—show some sort of consciousness or self-awareness? One might think so, at least in moments such as on the day when princess Diana died, or more recently, on that day in April 2013

P.A. Gloor (✉)
MIT Center for Collective Intelligence, Cambridge, MA, USA
e-mail: pgloor@mit.edu

A. Fronzetti Colladon • R. Pellegrini
Department of Enterprise Engineering, University of Rome Tor Vergata, Rome, Italy
e-mail: fronzetti.colladon@dii.uniroma2.it; romina.pe91@gmail.com

C.Z. Miller
Illinois Institute of Technology, Chicago, IL, USA
e-mail: cmille31@stuart.iit.edu

© Springer International Publishing Switzerland 2016
M.P. Zylka et al. (eds.), *Designing Networks for Innovation and Improvisation*,
Springer Proceedings in Complexity, DOI 10.1007/978-3-319-42697-6_13

when one of the authors was stuck at home in Cambridge while the Boston Marathon bomber was roaming at large in the neighborhood. In those intense moments, we feel maybe not "collectively intelligent" but certainly "collectively aware" or "collectively conscious." If we meet a stranger in those moments, we know what they are thinking, namely "it's so sad Diana died" or "where might the marathon bomber be hiding and hitting next." Moments like these motivate an informal definition of "organizational consciousness." It is analogous to the human body, where the brain is conscious of the toe, and will respond differently depending on whether a person hits her toe at the door or somebody else steps on her toe. Extending this metaphor, a "collectively conscious" organization will respond differently if somebody hits a member purposefully, or if a member hurts her/himself. Similarly to the neurons in the brain that are communicating through their synapses to create consciousness, humans communicate by interacting with each other verbally, through text, or other signals, either face-to-face or over long distance by phone or Internet.

To prove existence of consciousness on the individual level, Descartes famously stated "cogito ergo sum"—I think, therefore I exist. Extending this definition to an organization, "if the organization thinks and acts as one cohesive organism, it exists" and thus shows collective consciousness, defining organizational consciousness as common understanding of an organization's global context that allows the members of the organization to implicitly coordinate their activities and behaviors through communication.

As an example of a global level event, in the case of the Boston Marathon bomber, everybody in the Boston area was trying to stay abreast of the most recent developments on Twitter, Facebook, and the News and looking out for traces of the terrorists. On the organizational level, a well-oiled team of software developers working together closely face-to-face, using chat, or using e-mail trying to debug a jointly developed application also shows a high level of organizational consciousness, as they are able to coordinate their work with minimal use of words.

13.2   Coolhunting with the Six Honest Signals of Communication

Our aim is trying to make this implicit understanding more measurable, similarly to brain researchers, who measure individual levels of consciousness by attaching probes to individual neurons, tracking the electrical flow of current flowing through synapses between the neurons. In our work, we measure interaction among people through "coolhunting" in online media such as e-mail, Twitter, Facebook, and blog posts, applying a framework of "six honest signals of collaboration" to assess the level of global consciousness (Clark 2001).

13.2.1 Coolhunting Overview

We use the coolhunting approach (Gloor 2010). It distinguishes between three different sources of information: the crowd, the experts, and the swarm. The difference is explained well through the metaphor of coolhunting for a restaurant as a tourist in a foreign city. Following all other tourists will bring us to the places where all the tourists go; these restaurants will be crowded, full of other tourists, expensive, and not particularly good. This is what following the crowd gives us, as the crowd likes to follow well-trodden paths.

If we ask the concierge in our hotel for a recommendation, we will end up in a better restaurant, with better food, but it most likely will still be full of tourists and much more expensive. This is what following the advice of the expert brings us. The problem with experts is that they take kickbacks from the organizations whom they recommend, as they are paid to give advice, just like the rating agencies Moody's and Standard & Poor's, which get paid from the same companies and government whom they are supposed to assess, leading to serious conflict of interest.

We will find the best places to eat if we visit the places popular with the local residents. The hard part is trying to identify the locals on the street and in a crowded restaurant, as they are hard to distinguish from the tourists. We might get some hints by looking at their clothing and listening to their language. We call this the swarm, leading in our restaurant example to the best meal at the lowest price.

When doing coolhunting on social media, we need to make the same differentiation between crowd, experts, and swarm, based on the source. Twitter usually gives us the wisdom (and madness) of the crowd, blogs, and online newspapers give us the (paid) wisdom of the experts, while the swarm might be found among Wikipedia editors, in Facebook groups, and on subject-matter specific online forums. Obviously, the intrinsically motivated swarm will give us the best information quality. Tracking the right hashtags on Twitter might also lead us to the swarm for a certain topic.

13.2.2 The Six Honest Signals of Collaboration

To measure the impact of a topic on global consciousness, we use the "six honest signals of collaboration." They were originally defined for measuring collaboration within organizations by analyzing e-mail archives (Gloor 2015); they can be similarly applied to online social media. They are based on key social network analysis metrics (Wasserman and Faust 1994) and include two metrics each for structure, dynamics, and content of the network.

The two structural metrics are central leadership and balanced contribution. *Central leadership* measures betweenness centrality of a network, indicating how much the network is dominated by one or a few leaders. *Balanced contribution* measures, through contribution index (Gloor et al. 2003), how much members of a group are senders or receivers of information, and if the information is contributed by a small subset of the group, while the other group members are passive information consumers.

The two dynamics-based metrics are rotating leadership and responsiveness. *Rotating leadership* measures how much members of the network take turns in leadership by tracking oscillation in betweenness centrality. *Responsiveness* measures how quickly one actor responds to another one, for example, in Twitter how quickly a tweet is retweeted, or one person responds back to a tweet from somebody else, and how many nudges (pings) it takes.

The two content-based metrics are *honest language* and shared context. The more the language in tweets or online forums includes both very positive and very negative language, the more honest it is. *Shared context* measured how much a group is defining their own vocabulary, making up their own words and abbreviations.

These six "honest signals of collaboration" have been measured in online social media using the Condor tool (www.galaxyadvisors.com), which automatically collects Twitter, Facebook, Blog, and Wikipedia data and calculates the metrics.

13.3   Results

We will now describe three case studies of measuring collective awareness through Coolhunting on social media. *Francogeddon* was the event January 2015 when the Swiss National bank overnight removed the fixed binding between Euro and Swiss franc, leading to huge turbulences at the global exchange markets. We compare this event against one preplanned and well-orchestrated event, the launch of the Apple watch in Italy. Our third case is the turbulent months when Greece was teetering on the brink of bankruptcy and was pondering *Grexit*, the exit from the Eurozone.

13.3.1   *Francogeddon: Uncapping the Swiss Franc—A Signal of Global Consciousness?*

We start illustrating global consciousness by the example of Francogeddon. On January 15, 2015, financial markets were in turmoil. In a surprise move—later termed Francogeddon—the Swiss National Bank removed the artificial exchange rate of Swiss Franc 1.20 to the Euro, which it had set and defended by buying massive amounts of Euro and Dollars since September 6, 2011. Within hours, the exchange rate between Euro and Swiss Franc fluctuated from 1.20 Francs per Euro to 95 Swiss cents per Euro, leading to massive losses at stock markets around the world, forcing hedge funds into insolvency.

Such an unexpected event at the financial markets offers a unique natural experiment to measure global consciousness of financial markets. Using Condor, we collected the most recent 12,000 tweets containing the string "Swiss Franc," as well as

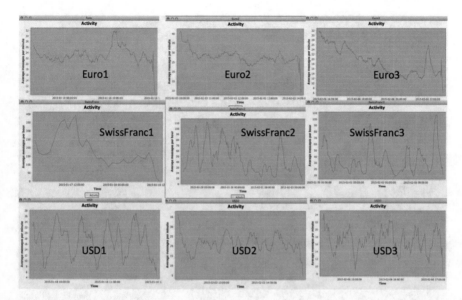

Fig. 13.1 Twitter activity after January 18, 2015 for search strings "Swiss Franc," "USD," and "Euro"

another 12,000 tweets each containing "Euro" and "USD" on January 18, when Francogeddon was still a major issue, and currencies were still fluctuating wildly. We repeated the data collection at two later points in time, on February 3 and February 6, 2015, when Francogeddon was over, and things had settled down. This nine-part dataset allows us to compare a moment of high public consciousness, when Francogeddon was at the top of everybody's minds involved in currency trading with a baseline of two later points in time when the event was over, and public consciousness should be low again.

The nine charts in Fig. 13.1 illustrate the activity of the tweeters on these 3 days. While the tweet activity about Euro and USD is about the same on all three sampling days (20–30 tweets per minute), tweet activity for Swiss Franc is about 200 tweets pro hour on January 18, dropping to 50 tweets per hour on February 3 and 6.

Figure 13.2 shows the network structure of the three currency twitter networks on January 18 and February 6. Each node is a person tweeting; a link is added between two nodes if one person is mentioned in the other's person tweet, or one person is retweeting the other person.

As Fig. 13.2 illustrates, the tweets about Swiss Franc on January 18 form a large connected component. The Euro network (which was more influenced by the Swiss Franc) shows a somewhat smaller connected component, while the USD tweet network is very little connected which tells us that the tweeters have nothing to do with each other. On February 6, all three tweet networks have similar structures of mostly unconnected tweets with the Euro still showing a somewhat larger connected component.

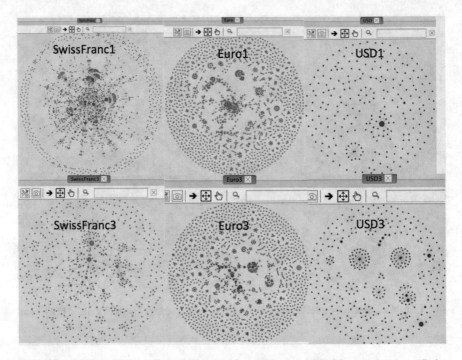

Fig. 13.2 Twitter network structure on January 18 and February 6, 2015 for search strings "Swiss Franc," "USD," and "Euro"

The six Word Clouds depicted in Fig. 13.3 show what people are tweeting about. While the sentiment about the Swiss Franc on January 18 is overarching negative (the darker the red of a keyword, the more negative its context), it is somewhat negative for the Euro tweets, and almost exclusively positive for the USD. The Swiss Franc tweets on February 6 are becoming more positive, but still mostly negative, as a lot of people in Eastern Europe, particularly in Poland, but also in Rumania and Austria, complain about taking out mortgages in Swiss Franc, which now ballooned against their local currency. A look at the USD tweets on both January 18 and February 6 shows that they mostly consist of retweets of items auctioned on eBay. This illustrates that the US tweeters do not care much about Francogeddon. Tweets about the Euro are somewhat negative, but the concerns—which are growing on February 6—are more about Mario Draghi and the possible Grexit, i.e., the exit of Greece from the Eurozone.

We calculated the six honest signals of communication for the nine datasets:

1. Group betweenness centrality (how centralized are the tweet networks)
2. Oscillation in group betweenness centrality (how much is the centrality of individual tweeters in the network changing over time, measured in 15 min intervals)
3. Average weighted variance in contribution index, i.e., how much are individual tweeters being retweeted over time

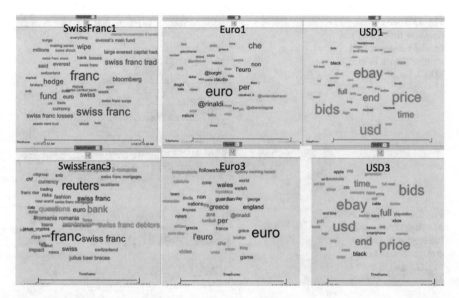

Fig. 13.3 Word cloud of tweets on January 18 and February 6, 2015 for search strings "Swiss Franc," "USD," and "Euro"

Fig. 13.4 Average emotionality of tweets containing search strings "Swiss Franc," "USD," and "Euro"

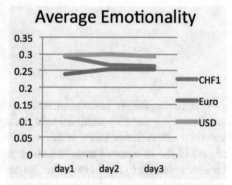

4. Average response time and nudges, which tells how long it takes for a tweet to be retweeted, and if people are mutually retweeting each other
5. Sentiment and emotionality, which show how positive and negative the tweets are
6. Complexity of language.

The charts below illustrate the changes over the three points in time in emotionality (see Fig. 13.4), average response time (ART) (see Fig. 13.5), and number of nudges per tweeter (see Fig. 13.6). For example, the response time (ART) drops considerably for USD from January 18 (day 1) to February 6 (day 3), while it goes up for Swiss Francs. This means things are cooling down for tweets about Swiss Francs, and it takes more time until they are retweeted.

Fig. 13.5 Average
response time (ART) of
tweeters using search
strings "Swiss Franc,"
"USD," and "Euro"

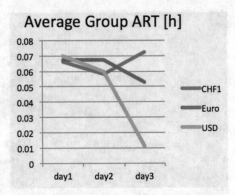

Fig. 13.6 Average number
of nudges (retweets) of
tweeters using search
strings "Swiss Franc,"
"USD," and "Euro"

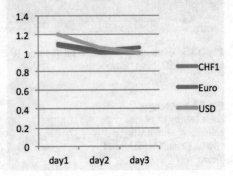

Comparing the six honest signals of communication for the three currencies, we see that even for this small sample, using the Mann–Whitney U-Test, tweeting behavior about Swiss Franc is different from tweeting about Euro and USD, with regard to the number of nudges as well as the variance between nudges until one tweeter responds to another tweeter. To put this in other words: comparing the three twitter networks about the three currencies over three points in time, there seems to be higher global consciousness by people tweeting about Swiss Franc compared to people tweeting about Euro and USD—a glimpse of global consciousness of currency traders related to Francogeddon?

13.3.2   Launch of the Apple Watch in Italy

The launch of the Apple Watch in Italy provides our second illustration of global consciousness. As previously noted, although the intrinsically motivated swarm will provide the best information quality, tracking the right hashtags on Twitter might lead to the swarm.

Data were collected in three different datasets between June 21 and July 11, which included the Apple Watch launch in Italy on June 26. One dataset was specifically for the Apple Watch, one for the collection of tweets on the smartwatch in general, and finally a dataset to collect tweets about a competitor of the Apple Watch, the LG Watch Urbane. Using Condor to analyze the three datasets, we were able to make comparisons between them based on measures of network structure, network dynamics, and network content. The tweet collection was restricted by geocode to Italy only.

We first observed the number of actors collected in each dataset. In the dataset "applewatch," there are 4970 actors. The number drops dramatically in the other two datasets: "smartwatch" has 907 actors and "LGwatch" 203 (Fig. 13.7). The theme Apple Watch involved a large number of Twitter users between June 25th and 26th corresponding to the delayed launch of the Apple Watch. In the other two datasets, there are far fewer actors, suggesting that the themes were not as "hot" or compelling as "applewatch."

In comparing the sentiment between the three datasets, there is relatively little variation. Overall, values were high as seen from the large number of green (positive) words in the word cloud. However, there is a decreased value for sentiment for "LGwatch" in the days prior to and shortly after the launch of the Apple watch. It is possible that conversations in the network were negatively affected by the arrival of the LG watch competitor. In the word clouds (see Fig. 13.8) for "smartwatch" and "LGwatch" many of the words that appear are related to the Apple watch, indicating how this event affected Twitter users that would not typically be tweeting about the Apple watch.

As in the example of the Francogeddon, we can observe radically different network structures and tweeting behaviors among the three sample data sets over the launch of the Apple watch in Italy, illustrating the different levels of collective awareness for the three different product launches.

13.3.3 The Greek Referendum

The Greek Referendum on July 5, 2015 provides the third illustration of global consciousness. From June 26 to July 7, the tweets with the search strings "GRoxi" and "GRnai" were collected. "Oxi-No" and "Nai-yes" stand for or against the austerity requirements of the EC, with a No-vote rejecting the austerity requirements of the EC, and risking a possible Grexit from the Eurozone. While the polls were predicting a close exit of the vote, the Greeks in fact soundly voted for "oxi," rejecting the austerity requirements.

Figure 13.9 illustrates the twitter network of the two hashtags. The "GRoxi" network is much denser, illustrating that in this case Twitter was a much better predictor of the exit of the vote than the official polls.

Figure 13.10 shows the word clouds for the two hashtags. As the Greeks were quite pessimistic in these times, it is not surprising that the words are mostly

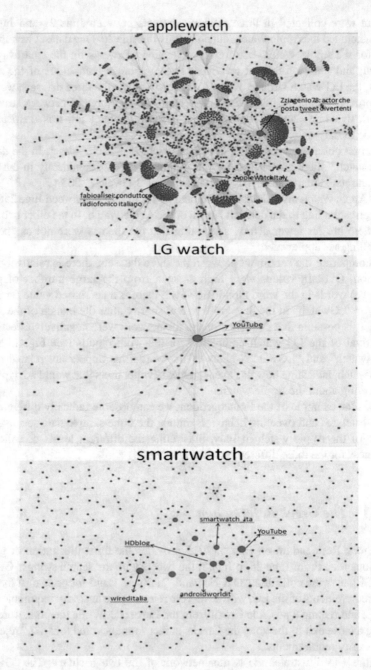

Fig. 13.7 Twitter network search structure for search strings "Apple Watch," "LG Watch," and "Smartwatch"

Fig. 13.8 Word clouds for search strings "Apple watch," "LG watch," and "Smartwatch"

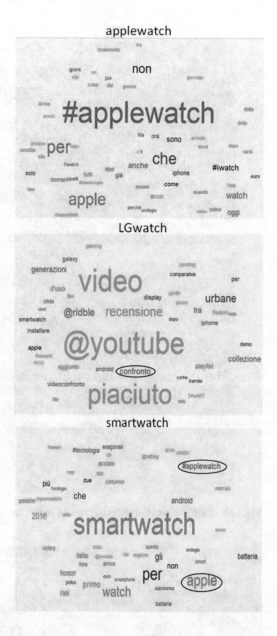

red—meaning they were used in a negative word context, as automatically measured by Condor's sentiment analysis tracking software. As an additional indicator of the outcome of the vote, the word cloud on "GRnai" includes "oxi" almost as large as "nai," with size indicating the frequency of the word. The GRoxi cloud does not prominently show "nai."

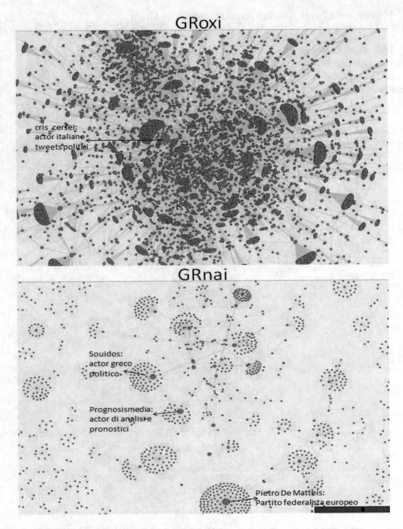

Fig. 13.9 Twitter network search structure for search strings "GRnai" and "GRoxi"

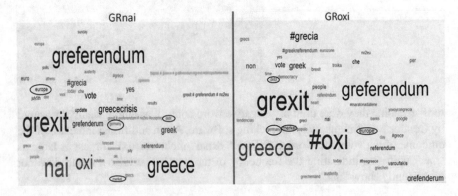

Fig. 13.10 Word clouds for search string "GRnai" and "GRoxi"

13.3.4   Comparative Analysis: Measuring Collective Awareness

In this section, we compare the results of the three different global events, comparing the magnitude of the signal for each of the events on the day when the event happened. Figure 13.11 shows sentiment, emotionality, complexity, average response time (ART), group betweenness centrality, and group degree centrality for the search term "Swiss Franc" on January 15, 2015, "Apple Watch" on June 26, 2015, and "GRoxi" on July 5, 2015.

We find the strongest signals of global awareness for Francogeddon. Emotionality is highest, and sentiment is most negative (sentiment is positive if its value is bigger than 0.5). This is quite surprising, as the Greek were quite unhappy with the austerity

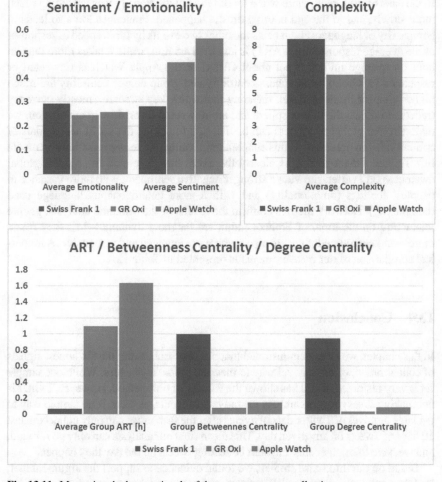

Fig. 13.11 Measuring the honest signals of the swarm to assess collective awareness

Table 13.1 Comparing the six honest signals of communication for the three events

	Francogeddon	Grexit	Apple watch
Average emotionality	0.293	0.244	0.258
Average sentiment	0.282	0.466	0.563
Average complexity	8.405	6.177	7.265
Average group ART [h]	0.066	1.099	1.636
Group betweenness centrality	0.998	0.077	0.146
Group degree centrality	0.943	0.035	0.087

measures introduced by the EC also, but it seems taking a large loss in 1 day leads to stronger expressions of frustration. Most of the negative tweets on Francogeddon came from currency traders and hedge fund managers who had to digest multibillion losses in 1 day, in some cases even leading to their bankruptcy and dissolution. In the case of the Greek, there was a lot of talk, but the tragedy was unfolding much more slowly, and in the end nothing drastic happened. Francogeddon also leads in complexity of language, and it beats the other two events by far in speed of response, as the average response time (ART) is less than a 5 min, while it takes more than an hour on average until a tweet about GRoxi or the Apple Watch is retweeted or responded to. Group betweenness centrality and group degree centrality are much higher also for Francogeddon, meaning that a few key tweeters, mostly currency traders, dominate the twitter sphere and are retweeted feverishly. In conclusion, we maintain that global awareness can be measured tracking the six honest signals of communication presented in this chapter, monitoring online conversations on media like Twitter. Francogeddon is clearly the event that generated the biggest global awareness on Twitter, showing a stronger negative sentiment, with more variance in peoples' feelings (emotionality) and with a more heterogeneous language used (complexity). Network metrics confirm the results of the semantic analysis: people interacting on the topic of Francogeddon are far more dynamic—i.e., they rotate more—and centralized; they are also much more engaged and responsive. A numerical comparison of our measurements is presented in Table 13.1.

13.4 Conclusion

In this chapter, we have demonstrated that our approach, using the six honest signals of collaboration, offers a novel way to measure global awareness. While our sample set is very restricted, it still has shown the validity of our method. However, Twitter is not making it easy for researchers to study such events, as there is no simple way to get large archives of Twitter data of events after the fact, as we can only collect the last 10 days of Tweets on any given day. This means that our analysis can only go forward, and we have to catch events and start collecting tweets on the day they happen.

Based on our three case studies, we found evidence to support the argument that, building on each other through tweets and retweets, actors are creating global awareness of key events. While all three events have left a recognizable footprint in global awareness, a sudden unexpected event with deep impact close to the bottom line such as Francogeddon leaves a much deeper impression in global awareness, than a carefully orchestrated marketing event such as the launch of the Apple watch, or a week-long litany of complaints about bad times such as the Greek vote on the Grexit. It seems that to leave a deep impact in global awareness, the surprise element is key.

References

Clark A (2001) Natural-born cyborgs? Springer, Berlin

Gloor PA (2010) Coolfarming: turn your great idea into the next big thing. AMACOM Div American Mgmt Assn.

Gloor PA (2015) What email reveals about your organization. Sloan Management Review, Winter

Gloor PA, Laubacher R, Dynes SB, Zhao Y (2003) Visualization of communication patterns in collaborative innovation networks-analysis of some w3c working groups. In: Proceedings of the twelfth international conference on information and knowledge management, ACM, pp 56–60

Sparrow B, Liu J, Wegner DM (2011) Google effects on memory: cognitive consequences of having information at our fingertips. Science 333(6043):776–778

Wasserman S, Faust K (1994) Social network analysis: methods and applications, vol 8. Cambridge University Press, Cambridge

Chapter 14
The Citizen IS the Journalist: Automatically Extracting News from the Swarm

João Marcos de Oliveira and Peter A. Gloor

14.1 Introduction

Wikipedia is one of the top ten websites, with more than 37 million articles in more than 250 languages. It lists approximately 27 million registered editors among whom 114,000 are listed as active contributors. Those users are spread all over the world, creating a 24 by 7 online community. This community quickly creates articles based on news coming from various news sources, with some articles even written by Wikipedians involved into the actual events (Iba et al. 2010).

Twitter is another website on the top ten list. It has 340 million active users every month; those users make on average 6000 tweets per second. Twitter also allows users to retweet and share links from external sources. Twitter provides various APIs that allow researchers to easily access its contents. In previous work (Petrovic et al. 2013), the author shows that Twitter includes information from all the main breaking news related to major online journals.

Earlier studies found that Wikipedia is in some cases faster than conventional news channels (Becker et al. 2011). These observations formed the foundation of the WikiPulse project (Futterer et al. 2013) and prompted (Fuehres et al. 2012) to propose the use of Wikipedia content to find "latest trends based on the analysis of recent edits on Wikipedia articles." Other studies had found that Twitter has become a popular news channel. The *SwarmPulse* project introduced in this chapter combines these ideas by generating latest news based on Wikipedia article edits and the most recent tweets, presenting them in a user friendly news format.

J.M. de Oliveira
Federal University of Juiz de Fora, Juiz de Fora, Minas Gerais, Brazil
e-mail: jmarcosdo@gmail.com

P.A. Gloor (✉)
MIT Center for Collective Intelligence, Cambridge, MA, USA
e-mail: pgloor@mit.edu

© Springer International Publishing Switzerland 2016   141
M.P. Zylka et al. (eds.), *Designing Networks for Innovation and Improvisation*,
Springer Proceedings in Complexity, DOI 10.1007/978-3-319-42697-6_14

The main contributions of this chapter are the SwarmPulse algorithm, which combines Twitter and Wikipedia to generate news automatically, the description of a first implementation, and an algorithm for measuring the accuracy of SwarmPulse.[1]

14.2   Related Work

In earlier research studying the collaborative behavior of Wikipedia editors, Bayer et al. (2011) found that unlike just many eyes having a look at an article, the experience of the editors is important—they should have worked on many other articles for the quality of their articles to be good. It was also found that a high number of editorial events contribute positively to a page's quality. In other earlier work (Becker et al. 2011), it was found that entertainment and sports news appeared on average about 2 h earlier on Wikipedia than on *CNN* and *Reuters* online. *Wikirage*, another Wikipedia-based news system, tracks the pages in Wikipedia which are receiving the most edits over various periods of time (Wood 2013). While this site does a good job collecting the edits, it does not process the results further and as evidenced in Bayer et al. work (2011), edits alone are not enough to justify newsworthiness. Nevertheless, Wikirage delivers a good benchmark to validate against the results of our news generation algorithm.

Twitter has also been used as a news detection tool; many research projects have proven its value to find relations between its data and real world events. Sakaki et al. (2010) uses tweets to build an earthquake detection system based on the frequency of tweets and hashtags in specific locations. Another project using twitter to find breaking news events is discussed in Petrovic et al. (2013); in this work, the authors build a First Story Detection (FSD) system using the tweets retrieved from specific users employing hashtags like "#breakNews," "#News," and so on, subsequently ranking and clustering those tweets into groups. The problem with this approach is the limitation by the number of users and that the system loses most of the comments that come from users outside of that list. *Twitterstand* is another application that uses twitter to identify breaking news. It increases its accuracy by only showing trusted sources. Also, in other research, it was found that although most of the breaking news can be found on twitter, twitter users mostly are not creating or contributing to news but comment on them (Subašić and Berendt 2011).

Another system using Wikipedia and Twitter for breaking news detection can be found in Osborne et al. (2012). In this system the authors analyze the use of Wikipedia as a possible filter for news extracted on Twitter. Wikipedia news detection is done by counting the number of views of a Wikipedia page in a time period then analyzing this data to identify the increase in page views. A similar analysis is made on Twitter. As a result, it was found that Wikipedia seems to lagging behind Twitter by about 2 h. In our system, we analyze the Wikipedia text information without looking at the frequencies of edits and news. While most of the pages identified by the system had an increase in the number of changes in the time period, we are using Twitter as our news filter.

---

[1] The reader can try out the prototype version of SwarmPulse at swarmpulse.galaxyadvisors.com.

14.3   SwarmPulse Algorithm

In order to create our SwarmPulse newsreader using data from Twitter and Wikipedia, we follow the steps as shown in Fig. 14.1.

1. Extract data from Wikipedia
2. Search recent tweets
3. Test their newsworthiness
4. Rank the article
5. Display the news

We will now describe these steps in detail.

14.3.1   Extract Data from Wikipedia

The first step to build our newsreader is to extract the most recently edited articles from Wikipedia using Mediawiki. Mediawiki is another project under the umbrella of the Wikipedia Organization. It gives access to the Wikipedia metadata. With this API, it is possible to request the links, categories, editors, and even the edit history of an article. Mediawiki also provides specific queries to collect the most recently edited articles from Wikipedia, which is exactly what we need. Once we have collected those articles, we have a list of potential news candidates.

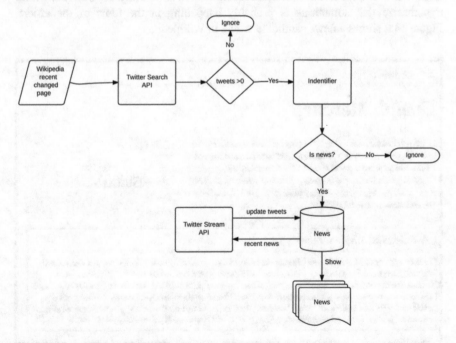

Fig. 14.1 SwarmPulse algorithm

To identify breaking news events, some studies suggest to analyze the connection between editors or the frequency of edits in the article (Ciglan and Nørvåg 2010). Although those methods had been proven to work (Futterer et al. 2013), SwarmPulse gives this task to the end users, by using Twitter as a first news filter.

14.3.2   Search Recent Tweets

The next step is to combine the Wikipedia article with the tweets. The Twitter API allows developers to have access to some of its data. In this project, we use two APIs: The Search API and the Streaming API. As its name suggests, the search API is used to make specific searches for tweets. It enables search for specific words, users, dates, and so on. However, it has some limitations. The main current limitation comes from Twitter's API in that it allows only 450 queries per 15 min window, with a limit of 100 results per query. In our implementation, we had to work around this restriction to ensure that it did not block processing of the data and put a limit on the number of Wikipedia pages the system can evaluate. The Streaming API gives access to the most recent tweets, it does not have a tweets limit. It depends of how fast one can process the data.

Now that we have explained how the Twitter API works we can continues with the news filter. Using the news candidates that we found using the Mediawiki, we can search for recent tweets using the Twitter Search API. For each news candidate from Wikipedia, we search for related recent tweets. If we find a keyword match, we hypothesize that something is probably happening in the topic of the article. Figure 14.2 shows a news candidate article in Wikipedia.

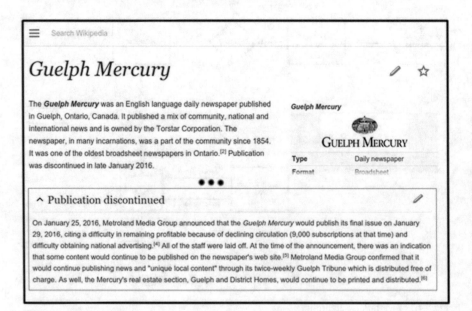

Fig. 14.2 Wikipedia news example

Table 14.1 Categories and subcategories

Categories	Subcategories
Business	Companies
Entertainment	Games, movies, music
People	Actors–actresses, business people, players, politicians, singers, writers
Sports	Sports, baseball, basketball, football
World	Africa, Asia, Europe, North America, South America

Fig. 14.3 Percentage of each news category in SwarmPulse

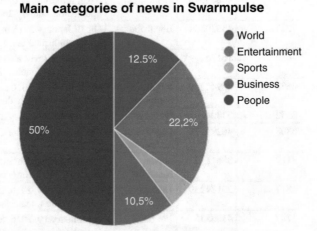

14.3.3 Identify Newsworthiness

For the news identification algorithm, we take advantage of the Wikipedians' writing style. Wikipedia by definition will give a history of events that implies telling when important events happened or are going to happen. So, using this information we can use a natural language processing algorithm to identify dates in the article and match those dates with sentences or paragraphs. In this step, we also process other data from the Wikipedia page. For instance, we collect the Wikipedia categories for a page and then use specific tags to map those categories into 5 categories and 19 subcategories (see Table 14.1). If the date is equal or close to the current date, we have found an event related to the present one where we already have tweets about it. Figure 14.3 shows the distribution of different categories in SwarmPulse.

14.3.4 Rank the Article

Finally, we need to rank the articles found in step 3. Again, we leave that task to the Twitter users. Using the Twitter Streaming API to collect the most recent tweets about the most recent events identified in the previous step, we then add those

Table 14.2 Top news on 2016-01-20

Total tweets	Page ID	Title	News
127,917	2941511	Sarah Palin	On January 19, 2016, palin endorsed Donald Trump's campaign to become preside…
28,412	175537	Netflix	Reedhastings admitted that netflix's china expansion could take "many years" on …
18,181	536880	Glenn Frey	Souther [30], Jack Tempchin [31], Irvingazoff [32], Lindaronstadt [33], Don Felder [34], …
16,137	40884573	@midnight	Note: minimum of three wins or five appearances, updated on January 26, 2016
7227	43630864	Jihadi John	On 19 January 2016 in the isil magazine dabiq, the group confirmed that Emwazi …
4682	16175	Jennifer Lopez	Ryanseacrest will also produce the series [176]. It premiered on January 7\
4633	28944259	Jorge Sampaoli	On 19 January, 2016 sampaoli and the chilean federation mutually decided to …
3952	33804	Wellington	Retrieved 19 January 2016
3808	46299779	Legends of Tomorrow	The series airs on the CW and premiered on January 21, 2016
2119	3861139	Steven Naismith	*Senior club appearances and goals counted for the domestic league only and correct…
1817	23184259	4Minute	On January 20, 2016, it was announced that the group would release their seventh mini album …
1787	4106856	Jim Schwartz	On January 19, 2016, Schwartz was hired by the Philadelphia Eagles to be their defensive coordinator [23].
1124	26903	Solar system	On January 20, 2016 astronomers at the California institute of technology announced a possible ninth planet…
1063	24831215	Zika virus	On 15 January 2016, CDC issued a level 2 travel alert for people traveling to regions and…

tweets to the article, ranking the news by the number of associated tweets. Table 14.2 shows the top ranked news on January 20, 2016; we can see in this table that the Sarah Palin endorsement of Donald Trump's presidential campaign was the news with the highest reflection on Twitter. Twitter also gives more information about the articles such as the reactions of people and links to other news sources.

14.3.5 Display the News

Once we have collected all this information, we display it on a specific website, with the event as a news header, including part of the Wikipedia page for more information as well as the tweets to allow users to search for more links and to look at reactions about the topic.

14.4   Data Analysis

In order to measure the performance of the news identification algorithm, we compared the news found in two different ways: comparing the news with the RSS feed from CNN and Reuters online and using the Google news page to get news for a specific search term, obtaining its results as RSS feeds. The main challenge is to match two news items reported from two different websites because these two news items might be about the same events using completely different words. To overcome this obstacle, we developed a keyword-based matching heuristic (see Fig. 14.4).

We first remove all stopwords from the RSS feeds and the news found in SwarmPulse. Then we process the content of each news item, creating two list of keywords from the Wikipedia article, one for the title and the other for the content. We repeat this process for the CNN, Reuters, and Google News RSS feeds. The news items from Wikipedia are then compared with the news items from the RSS feed using the keywords. A match is expressed in terms of *match strength*—a fraction

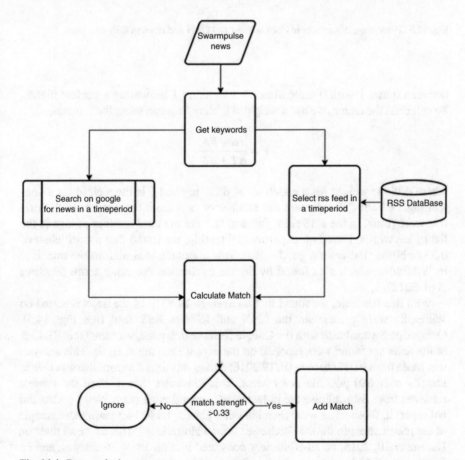

Fig. 14.4 Data analysis process

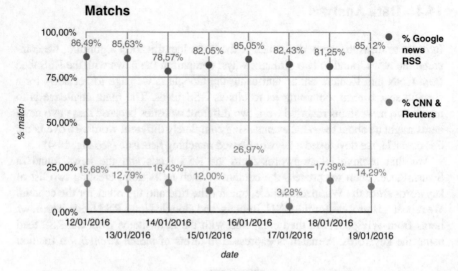

Fig. 14.5 Percentage of matches in Google news and CNN and Reuters RSS data feed

between 0 and 1 with 0 indicating no match and 1 indicating a perfect match. To calculate the match, we use a weighted arithmetic mean using the formula,

$$F = \frac{\alpha t + \beta c}{\alpha T + \beta C}$$

where α is the weight for a match on a "title" tag and β is the weight for a content tag. "c" and "t" represent the number of tags from the content or the title that were found in the RSS feed. "T" and "C" are the total number of tags in the list of keywords. Based on experimental testing, we found that a coefficient of 0.33 or higher indicates a good match. This approach is similar to the one used in WikiPulse which also found by human evaluation the same score (Fuehres et al. 2012).

With this heuristic, we found that on average 14.85 % of the news reported on Wikipedia were present on the CNN and Reuters RSS feed (see Fig. 14.5). Comparing SwarmPulse with the Google News search page, we found that 83.32 % of the news we found were reported on the days we ran the analysis. This analysis was made from 01/12/2016 to 01/19/2016; during this time SwarmPulse was able to identify over 680 possible news items. This illustrates that most of the content retrieved from SwarmPulse are in fact news, and although many News outlets did not report it, those news were found in other places. Table 14.3 shows an example of the match strength for the Wikipage "Daniel Holtzclaw," with the news that "on December 10, 2015, an all-white jury convicted him on 18 of 36 charges, and on January 21, 2016 he was sentenced to 263 years in prison."

Table 14.3 Match strength example

News source	Best news match	Match strength
CNN	Danielhotlzclaw the ex Oklahoma city officer convicted of rape …	0.425
Reuters	To access the newsletter click on the link http share thomsonreuters …	0.225
Google news	The latest ex officer convicted of rape won t get new trial …	0.450

14.5 Limitations

Although a user generated news portal opens up many new opportunities, it will have limitations with regard to the content generated. Wikipedia's main focus is on global or national events, which makes its data less useful for local news, except for local news of national interest. Another key question is the trustworthiness of the Wikipedia pages. While Wikipedia is geared towards discovering fake news quickly, for a limited amount of time a fake Wikipedia page can show up once we have some tweets covering the subject (there was for instance a case where for a short period of time the Ebola page claimed that Ebola was caused by gays). A possible next step might be to test the resulting SwarmPulse news articles with human readers to determine its usefulness. It may be necessary to have a human editor edit the cacophony of such a unique combination of text to present a "journalist's" coherent written story.

Another open issue is the dependency of our system on external APIs, while Wikipedia is free and open source, Twitter is run by a commercial company which over the past years has repeatedly changed availability and terms of use of their Twitter stream.

14.6 Future Work and Conclusions

The extraction of news from Twitter and Wikipedia opens the door to many exciting possibilities. Among those possibilities are the connection between articles, the sentiment analysis of the tweets, data visualizations, and others.

We believe that this information can be used in many different fields. For instance, it can be used by conventional News media to identify breaking news through data collected from social media and to gauge the relevance of news to users. Also, due to the information from the tweets, we have direct links to different sources of information that can be applied to create an index to news articles.

We also believe that data visualization is a big opportunity to create cybermaps that represent the network structure between different news items, giving us a first impression of how events might be connected with each other.

The main contribution of our work is a novel news reader that combines Wikipedia articles and Twitter data. It opens new windows of opportunities to different types of analysis, leveraging the power of the "citizen journalist" as a trusted provider of late breaking news.

References

Bayer T, Ford H, Tar D, Romanesco (2011) Quantifying quality collaboration patterns, systemic bias, POV pushing, the impact of news events, and editors' reputation. http://en.Wikipedia.org/wiki/Wikipedia:Wikipedia_Signpost/2011-11-28/Recent_research

Becker H, Naaman M, Gravano L (2011) Beyond trending topics: real-world event identification on twitter. In: Proceedings of fifth international AAAI conference on weblogs and social media, AAAI

Ciglan M, Nørvåg K (2010) WikiPop: personalized event detection system based on wikipedia page view statistics. In: Proceedings of the 19th ACM international conference on information and knowledge management, ACM, New York. doi:10.1145/1871437.1871769

Fuehres H, Gloor PA, Henninger M, Kleeb R, Nemoto K (2012) Galaxysearch—discovering the knowledge of many by using wikipedia as a Meta-Searchindex. Paper presented at collective intelligence conference, 2012 (arXiv:1204.2991). https://arxiv.org/pdf/1204.3375.pdf

Futterer T, Gloor PA, Malhotra T, Mfula H, Packmohr K, Schultheiss S (2013) WikiPulse—a news-portal based on wikipedia. Paper presented at COINs13 conference, Chile, 2013 (arxiv:1308.1028)

Iba T, Nemoto K, Peters B, Gloor PA (2010) Analyzing the creative editing behavior of wikipedia editors: through dynamic social network analysis. Procedia Soc Behav Sci 2:6441–6456. doi:http://dx.doi.org/10.1016/j.sbspro.2010.04.054

Osborne M, Petrovic S, McCreadie R, Macdonald C, Ounis I (2012) Bieber no more: first story detection using Twitter and Wikipedia. In: Proceedings of the SIGIR workshop in time-aware information access. Association for Computing Machinery

Petrovic S, Osborne M, McCreadie R, Macdonald C, Ounis I, Shrimpton L (2013) Can Twitter replace newswire for breaking news? In: Proceedings of the international AAAI conference on web and social media

Sakaki T, Okazaki M, Matsuo Y (2010) Earthquake shakes Twitter users: real-time event detection by social sensors. In: Proceedings of the 19th international conference on world wide web

Subašić I, Berendt B (2011) Peddling or creating? Investigating the role of twitter in news reporting. In: Clough P, Foley C, Gurrin C, Jones G, Kraaij W, Lee H, Mudoch V (eds) Advances in information retrieval. Springer, Berlin

Wood C (2013) Wikirage. http://www.wikirage.com/topedits/

Chapter 15
Growth Hacking: Exploring the Meaning of an Internet-Born Digital Marketing Buzzword

Timo Herttua, Elisa Jakob, Sabrina Nave, Rambabu Gupta, and Matthäus P. Zylka

15.1 Introduction

Internet startup companies are looking for ways to grow their business. Individuals involved in the Internet industry discuss and share experiences on how to best accomplish the goal of rapid growth. One of these people is Sean Ellis, affiliated with such growth companies as Dropbox, LogMeIn, and Qualaroo. He is credited for coining the term growth hacking, which has recently gained attention in the Internet startup industry (Ellis 2010; Holiday 2013). Missing an official definition, the term has caused discussions in the industry regarding its usefulness and boundaries.

The definition of growth hacking has many implications on business. After a broad search through the Internet blogs and articles, there is a huge misunderstanding of the term growth hacking. As many blog articles try to handle the term growth hacking, there is no scientific definition of the term.

The goal of this chapter is to stimulate the academic discussion about growth hacking and differentiate it from the traditional marketing strategies viral marketing and guerilla marketing. Our research question, then, is: *What is the definition of growth hacking and what are the main characteristics of growth hacking?*

To answer this research question, we designed a multi-method study (Mingers and Brocklesby 1997; Mingers 2001, 2003) that consists of two parts. We conducted

T. Herttua
Aalto University, Helsinki, Finland

E. Jakob • S. Nave • R. Gupta
University of Bamberg, Bamberg, Germany

M.P. Zylka (✉)
Department of Information Systems & Social Networks,
University of Bamberg, Bamberg, Germany
e-mail: matthaeus.zylka@uni-bamberg.de

© Springer International Publishing Switzerland 2016   151
M.P. Zylka et al. (eds.), *Designing Networks for Innovation and Improvisation*,
Springer Proceedings in Complexity, DOI 10.1007/978-3-319-42697-6_15

expert informant interviews with 12 startup founders and a case study in the qualitative first part of our study. In the quantitative second part of our study, we conducted a topic model analysis of 1.7 million Tweets.

This study's contribution to research and practice is threefold. First, we suggest a definition built from rigor scientific methods. Second, we explain the main characteristics of growth hacking. Third, we provide a growth hacking process model that may be used by researcher and practitioners alike.

The remainder of this chapter is organized as follows. The next section outlines a theoretical background on viral marketing and guerilla marketing, marketing strategy concepts that are related with growth hacking. In Sect. 15.3, we depict our course of action, followed by a section that presents the research results. This chapter concludes with a discussion of the results, limitations, and implications for future research.

15.2 Background

When discussing the meaning of a term, — in our case growth hacking — understanding how meaning is assigned to a term is of importance. Langlois (2014) writes about the substance of meaning as difficult to pinpoint, since meaningfulness is partially about personal experience and about processes of orientation of the individual. Therefore, there is a wide scope of interpretation concerning the term growth hacking. Many Internet articles define it as a digital way of leading marketing into new directions. Because growth hacking can be classified as a marketing strategy (Holiday 2012), we want to establish its difference from related concepts, viral marketing, and guerrilla marketing. They have similar characteristics, including the use of communication through social media channels to amplify the effect of campaigns.

The term guerrilla marketing was coined by J.C. Levinson (1998). It is defined as an achievement of goals, "such as profits and joy, with unconventional methods, such as investing energy instead of money." (Levinson 1998). Hutter and Hoffmann (2011) provide an up-to-date definition for guerilla marketing. Guerilla marketing summarizes unconventional low-priced marketing campaigns that draw the attention of a large number of clients (Hutter and Hoffmann 2011). Surprise, diffusion and low cost effect are the main drivers of guerilla marketing (Hutter and Hoffmann 2011).

Viral marketing is a guerilla marketing instrument that mainly focus on diffusion (Hutter and Hoffmann 2011). Viral marketing can be explained with information, which spreads like a virus (Jurvetson 2000). Social media channels are the main platforms to perform this marketing strategy. It works with word-of-mouth advertising in social networks. If it is adopted in the network, its diffusion will be viral (Leskovec et al. 2007). This strategy is very useful for niche products, as they need a certain critical mass (Jurvetson 2000).

15.3 Methodology

15.3.1 Part I: Qualitative Analysis

For our qualitative research, we referred primarily to Creswell (2013), which deals with the topics qualitative inquiry and research design.

15.3.1.1 Interview Analysis

We followed a grounded theory (GT) approach, originally developed by Glaser and Strauss (1967). GT is based on several steps, like data collection and analysis, verifying and presenting or publishing the results (Corbin and Strauss 2015). GT is also defined as either the creation or discovering a theory of a process (Creswell 2013; Corbin and Strauss 2015). GT is well suited for situations, where no previous scientific theory about the research subject exists (Creswell 2013). Since this was the case with our research subject growth hacking, we chose to use GT as the methodological basis of this part of our research. Analysis in GT is composed of three coding procedures, called open, axial, and selective coding (Corbin and Strauss 2015).

We interviewed 12 companies. Therefore, we used a semi-structured interview format for all interviews, but each interview also had room for more open-ended discussions (Myers 2013). The interviews were transcribed and transferred to Microsoft Excel in order to categorize them and create comparable results. Then, the interviews were manually coded, using the open coding approach (Corbin and Strauss 2015; Creswell 2013). After this step, we conducted axial coding (Corbin and Strauss 2015) by connecting similar categories (results from the open coding procedure). Finally, we wrote down all insights and discoveries out of the interview data into a summary, which corresponds to the selective coding method characterized by Corbin and Strauss (2015), and try to match them with the results from the analysis of the case study, explained in the following.

15.3.1.2 Case Study Research

A case study research is a qualitative approach in which exploration of either a real life case or cases over time in detail is needed to get a significant data collection through multiple sources of information (Creswell 2013). It is needed to analyze all existing information to get a meaningful overview and a wide range of qualitative materials (Creswell 2013).

We conduct a case study with one company to verify the collected and analyzed interview data. For the sequence of events in the case study, Creswell (2013) suggests creating a chronicle of the analyzing process. We develop a timetable to comprehend the whole case study process.

In the growth hacking process, we used the types of information according to Creswell (2013): observations, interviews, and audiovisual materials. In our study, as a way of gathering observations, we looked through the case company's social media channels. This analysis was done to find out the optimal channels and timing of sending out social media posts, in order to reach the widest possible audience, and to understand which content resonates best with the audience.

As a part of the case study, interviews with company representatives were conducted through the whole project. We had multiple discussions, in meetings or via E-Mail, where we planned the next steps of the case study, for example, regarding the planning of upcoming content.

15.3.2   Part II: Quantitative Analysis

We collected 1.7 million Tweets with the hashtag #growthhacking between May and October of 2015. In order to try to distil the Twitter discussions of growth hacking into groups that are manageable in size from a human perspective, machine learning, especially topic modeling, became an interesting candidate for analysis.

Blei (2012) describes topic models as "algorithms for discovering the main themes that pervade a large and otherwise unstructured collection of documents. […]. Topic modeling algorithms can be applied to massive collections of documents." (Blei 2012). Further, topic modeling can be conducted for short text scenarios such as Tweets (Hong and Davison 2010). We conducted topic modeling analysis (DiMaggio et al. 2013) with *Latent Dirichlet Allocation* (LDA) and *Hierarchical Dirichlet Process* (HDP) algorithms, utilizing an open source Python topic modeling library gensim (Rehurek 2010).

The challenge with using unsupervised clustering algorithms, including topic models such as LDA and HDP, is that the interpretation of results needs to be conducted by humans—the model itself does not mean anything specific. Since a noisy dataset produces a model where there is a lot of overlap between topics, one approach for making sense of the topics is to cluster the topics with an unsupervised clustering algorithm such as hierarchical clustering, which is defined from Grira et al. (2005) as "obtain a hierarchy of clusters, called dendrogram, that shows how the clusters are related to each other." We used the open source KNIME analytics platform to produce clusters and then observe which keywords seem to be related.

Since topic modeling algorithms proved to be problematic in noisy data, an alternative approach to quantitatively analyzing the contents of a dataset is to count the tweets with mentions of a certain search term in order to understand how relevant such keywords are within a corpus of text. Chew and Eysenbach (2010) have conducted a similar analysis, referring to it as infoveillance, a term used also by Eysenbach (2009) in another paper, referring to using open media intelligence techniques to closely monitor a discussion.

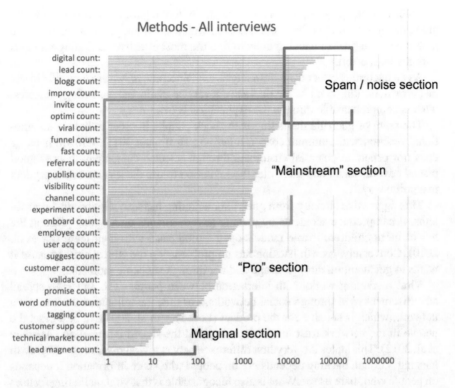

Fig. 15.1 Twitter mention counts for codes in interview codeset

When analyzing the counts of keywords, we looked at the results of the manual codes from the interviews we analyzed and then used the codes as keywords to search for (see Fig. 15.1 for examples of keywords and related mention counts).

15.4   Results

15.4.1   Part I: Qualitative Analysis

15.4.1.1   Interview Results

In the practitioner interviews, interviewees saw growth hacking differing from traditional marketing in that it leverages technology and, therefore, has better measurability. The interviewees used terms such as "analytics-driven," "metrics-driven," "tech-driven," and "data-informed" to describe the technical aspect of growth hacking. Although growth hacking is often done using digital channels, the interviewees said that it could be done related to non-digital products and services as well.

The word "hacking" seems to relate to the idea of looking for cost-effective ways to grow, being able to effectively operate even with a small budget—one interviewee mentioned that growth hacking aims to find the most effective marketing methods with the least effort.

As an upcoming effect from the analysis explained above, we are now able to say that growth hacking can't be put on a same level like traditional marketing strategies, what is comprehensible through the explanation below.

The point of guerrilla marketing is to shock people, in order to induce an emotional reaction with minimal costs (Hutter and Hoffmann 2011). Growth hacking does not penetrate potential customers through selected campaigns. It uses good placed advertisement and thereafter the invitation to use the service or the product in a polite way.

This aggressive strategy from guerrilla marketing has also a lot of ethical problems. It can produce accidents by placing advertisements on a wrong place at the bus or harm children, if role models open a toothpaste with their teeth (Ay et al. 2010). Conversely, growth hacking has no problems on the ethical side, because it wants to get attention through usage and not through shocking moments.

Viral marketing works with interaction between people. The aim is to spread advertisement *viral* through social networks. For this, the influential actor needs a network, which is suitable for the product (Leskovec et al. 2007). In best case, the people in the network trust the actor and spread the shared information (Aghdaie et al. 2012). This states the very first difference between viral marketing and growth hacking. Growth hacking depends not on people who share information it depends on people who share usage. What is also recognizable is that viral marketing focuses on special campaigns, for example, a special video (Aghdaie et al. 2012). Therefore, it is limited for the time period in which the campaign starts and the information is spread. Growth hacking, however, happens all the time. It is not limited in time and is not attached to special campaigns.

All in all, it can be said that growth hacking has indeed similarities to other marketing strategies, but the differences prevail. One general difference between growth hacking and traditional marketing strategies is that a growth hacker needs technical background, because one main task of a growth hacker is developing.

15.4.1.2 Case Study Results

The representative of the case company, which is a sharing economy startup, accepted our proposal to implement Google Analytics conversion events and then to observe the relative effectiveness of each channel from the perspective of how they drove revenue. The major challenge in the process was that the company has outsourced its technology team. In the end, the process of implementing the events on the site and in the product took 2 months from acceptance of the action plan to deployment of the analytics events.

Out of all of the six existing channels with visits attributed to them by Google Analytics, three brought any conversions. Fifty-four percent of the overall unique visitor volume came from channels that had no conversions, and the converting channels represented 46 % of the traffic. Thus, the conversion for only the converting channels was 4.47 %, whereas the overall conversion rate was just 2.1 %. The analysis yielded an insight that organic search and directory listings might be turned into marketing activities, for instance, by using paid search and paid directory listings. Experiments with these approaches have already brought promising results.

In summary, this case study strengthened the view that technical skills are core to the fast feedback cycle that is a cornerstone of growth hacking, and that end-to-end measuring of the user lifecycle is required in order to realistically inform which activities are value adding in terms of generating growth, which ones are creating traffic but no real revenue growth, and whether there are any currently small but promising channels to further invest into.

15.4.2 Part II: Quantitative Analysis Results

We started interpreting the results of the LDA and HDP topic modeling by using hierarchical clustering. We tried a number of different cluster counts for parameters to the hierarchical clustering module and ended up with using 25 clusters. The top six clusters were further inspected, and their keywords analyzed.

As a result, we created an Excel sheet with the top six clusters and their keywords with interesting keywords highlighted. We used these keywords to inform our next analysis step, the count of methods in Twitter data.

When conducting the practitioner interviews described above, we were asked by the interviewees about a list of methods related to growth hacking. As an alternative approach to describing growth hacking by observing methods related to it, we turned to counting mentions of important keywords that we have picked up from the qualitative analysis.

Certain search terms, such as *content*, *guide*, and *ebook*—all relevant codes from our interview analyses—were mentioned so many times that one can interpret them at least partially as "spam," whereas others were mentioned only a few times in the whole 1.7 M tweet dataset, making them marginal—mentioned but not necessarily characteristic of growth hacking. Keywords with mentions in the 1000–9999 range could be characterized as *mainstream*, and keywords with mentions in the 100–999 range could be characterized as *professional* keywords meaning they are not popular in the mainstream discussion, but still important to know for a professional interested in growth hacking.

When grouping keywords with 1–10k tweets, they seem to revolve around, on one hand, different channels of distribution, content, and related tactics, and on another, around the strategic and abstract as well as product and messaging related aspects. All in all, the keywords are a mixture of traditional, well-recognized concepts like conversion funnels, lead generation, audience, and message—but also add into the mix the latest trends of content marketing, infographics, podcasts, and social media.

15.4.3 Growth Hacking Process Diagram

After collecting all our results, we have analyzed multiple perspectives to the term growth hacking. In order to visually present our understanding of the growth hacking process, we have drawn a process diagram. This process refers to the results, which we made during this study.

The first step at all is to *analyze the actual situation* of the company and the product with data-dependent methods. If there already are social media performances or other available data, analyze it and ask what is needed to find out from this data. For this step, it is very useful to measure data with data analytic tools with more than one significance.

The next step is the *product optimization*. For this step, design plays a very important role. One difference between growth hacking and traditional marketing strategies is that growth hacking makes a product people want and not make the people want the product. So now the results from the data analytics have to turn into ideas, which fit to the product. From this arises a general sequence of steps, which can be developed to best practices in product optimization.

After the product seems to be perfect, it is necessary to perform *A/B tests* with at least two versions of a product. Before test phase begins, it is important to determine the goals from the tests. There should be only small number of goals to ensure reachability. The tests should be performed only based on one changing variable, because otherwise tests can end in complicated results which can't be worked out. The version with the better statistics wins.

After A/B testing, the company and the tested product should be ready for the proper *hack* step. This step depends on the company's aims and goals and also on learning effects during the whole process. For the hack, the first thing to clarify is the content and the design. After that, a strategy is needed to come up with the first tactical growth hack. For the whole growth hack, it is important that nearly everything is measureable, analyzable, and implementable. In the hack step, it is not necessary to acquire new users, but it is necessary to convince the existing users so that they won't churn. At the end of the day, it is important that users understand why the product is more valuable than other similar products, and why they should trust and be convinced from the company.

Growth hacking is a permanent process, and it is needed to repeat this process again and again to be a successful growth hacker (Fig. 15.2).

15.5 Implications and Limitations

Now that we have paved the way into looking at different perspectives of growth hacking, a subsequent study on the topic would benefit from choosing a narrower perspective and try to, for instance, replicate some of the results from this study.

A structured or semi-structured survey of a wide range of practitioners would also be interesting, for instance about the attitudes or capabilities towards the

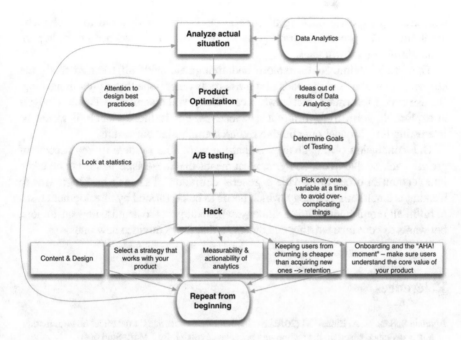

Fig. 15.2 Growth hacking process diagram

methods mentioned in this study. Broader quantitative research may emphasize our results and would be able to look at special points in more detail, especially in collaboration with other companies. A very big lack in research is also that there is not a sequence on techniques to apply on growth hacking. It would be interesting if the applied techniques have always to differ depending on companies with product or service.

Our study is limited in several ways. We have used 1.7M Tweets, 12 interviewees from 11 companies, and one case company, so although not necessarily statistically significant, this study does represent a wide variety of data sources. We cannot make bold claims about having much reproducibility of this study—someone replicating this study will most likely get a slightly different result by having different interviewees and a different dataset to work on.

15.6 Conclusion

This study produced a variety of different types of data, which can be used together to triangulate an understanding about what growth hacking is.

The interview analysis gave a perspective to marketing practitioners' views on the subject. The Twitter analysis helped pick out themes that resonate across the whole corpus of 1.7M tweets, and give an overview of what topics are part of the mainstream of growth hacking discussion on Twitter, as opposed to more niche discussions. The process diagram gives an overview what tasks need to be kept in mind during the growth hack.

This study attempted to explore and triangulate with different methods the essence, definition, and methods of a grass-roots practitioner term growth hacking. We use expert informant interviews, a case study, and quantitative Twitter analysis to explore the term from different perspectives. For further research, it would be interesting how broader quantitative research emphasize our results.

One conclusion of this chapter is that we were able to identify key aspects of growth hacking due to our analysis, what means that these aspects need to be taken into account in order to provide a general definition of growth hacking: Growth hacking is a marketing strategy, which needs to be performed by development tasks to fulfill all requirements. It is a strategy, which aims not to acquire new customers, but wants to convince existing customers so that they convince new ones.

References

Aghdaie S, Sanayei A, Etebari M (2012) Evaluation of the consumers' trust effect on viral marketing acceptance based on the technology acceptance model. Int J Mark Stud 6(4)

Ay C, Pinar A, Sinan N (2010) Guerrilla marketing communication tools and ethical problems in guerilla advertising. Am J Econ Bus Adm 2(3):280

Blei DM (2012) Probabilistic topic models. Commun ACM 55(4):77–84

Chew C, Eysenbach G (2010) Pandemics in the age of twitter. PLoS One 5(11):e14118

Corbin JM, Strauss A (2015) Basics of qualitative research: techniques and procedures for developing grounded theory, 4th revised edn. Sage, Thousand Oaks

Creswell JW (2013) Qualitative inquiry and research design: choosing among five approaches. Sage, Thousand Oaks

DiMaggio P, Nag M, Blei D (2013) Exploiting affinities between topic modeling and the sociological perspective on culture: application to newspaper coverage of U.S. government arts funding. Poetics 41(6):570–606

Ellis S (2010) Find a growth hacker for your startup. http://www.startup-marketing.com/where-are-all-the-growth-hackers/. Accessed 1 Apr 2015

Eysenbach G (2009) Infodemiology and infoveillance: framework for an emerging set of public health informatics methods to analyze search, communication and publication behavior on the Internet. J Med Internet Res 11(1):e11

Glaser BG, Strauss A (1967) Discovery of grounded theory: strategies for qualitative research. Sociology Press, Mill Valley

Grira N, Crucianu M, Boujemaa N (2005) Unsupervised and semi-supervised clustering: a brief survey. In: A review of machine learning techniques for processing multimedia content, Report of the MUSCLE European Network of Excellence (6th Framework Programme)

Holiday R (2012) Everything is marketing: how growth hackers redefine the game. http://www.fastcompany.com/3003888/everything-marketing-how-growth-hackers-redefine-game. Accessed 12 Dec 2015

Holiday R (2013) The secret that defines marketing now. http://www.fastcompany.com/3013859/the-secret-that-defines-marketing-now. Accessed 11 Dec 2015

Hong L, Davison BD (2010) Empirical study of topic modeling in twitter. In: Proceedings of the 1st workshop on social media analytics (SOMA '10)

Hutter K, Hoffmann S (2011) Guerrilla marketing: the nature of the concept and propositions for further research. Asian J Mark 5(2):39–54

Jurvetson S (2000) What exactly is viral marketing? Red Herring 78:110–112

Langlois G (2014) Meaning in the age of social media. Palgrave Macmillan, New York

Leskovec J, Adamic LA, Huberman BA (2007) The dynamics of viral marketing. ACM Trans Web 1(1):1–39

Levinson JC (1998) Guerrilla marketing: secrets for making big profits from your small business. Houghton Mifflin Harcourt, Boston

Mingers J (2001) Combining IS research methods towards a pluralist methodology. Inf Syst Res 12(3):240–259

Mingers J (2003) The paucity of multimethod research: a review of the information systems literature. Inf Syst J 13:233–249

Mingers J, Brocklesby J (1997) Multimethodology: towards a framework for mixing methodologies. Omega 25(5):489–509

Myers MD (2013) Qualitative research in business & management. Sage, London

Rehurek R (2010) Software framework for topic modelling with large Corpora. In: Proceedings of the LREC 2010 workshop on new challenges for NLP frameworks, pp 45–50

Chapter 16
ADR Processes for Creating Strategic Networks for Social Issues: Dementia Projects

Makoto Okada, Yoichiro Igarashi, Hirokazu Harada, Masahiko Shoji, Takehito Tokuda, and Takashi Iba

16.1 Introduction

Contemporary social issues are intertwined in an extremely convoluted way. Developing workable solutions in these complex circumstances requires that all sectors of society envision the future and cooperate as joint stakeholders in creating that future.

The overall situation with respect to social issues is wide and ambiguous—it is not at all clear who these stakeholders that might inject new energy into the issues actually are. Essentially, we need to gather people from various sectors who can participate as future stakeholders and then start looking for sound solutions while carefully defining the issues that need to be resolved. Then, through collaboration, these stakeholders will work together to clearly identify the questions and issues that need to be addressed.

With the above in mind, we focused in this study on social issues related to dementia to develop our understanding of the innovation design process (Okada

M. Okada (✉) • Y. Igarashi • H. Harada
Fujitsu Laboratories Ltd., Kawasaki, Japan
e-mail: okadamkt@jp.fujitsu.com; y-igarashi@jp.fujitsu.com;
hirokazu.harada@jp.fujitsu.com

M. Shoji
GLOCOM: Center for Global Communications, International University of Japan,
Tokyo, Japan
e-mail: shoji@glocom.ac.jp

T. Tokuda
Dementia Friendship Club, Tokyo, Japan
e-mail: tokuda.takehito@gmail.com

T. Iba
Faculty of Policy Management, Keio University, Tokyo, Japan
e-mail: iba@sfc.keio.ac.jp

© Springer International Publishing Switzerland 2016
M.P. Zylka et al. (eds.), *Designing Networks for Innovation and Improvisation*,
Springer Proceedings in Complexity, DOI 10.1007/978-3-319-42697-6_16

et al. 2013). Our hypothesis is that dementia issue can function as a boundary object that brings together future stakeholders (Nonaka and Konno 2012; Star and Griesemer 1989). We endeavor to demonstrate this process through case studies.

We are also interested in applying technology road-mapping (TRM) tools for social issues. TRM tools are not limited to use as communication tools—they can also be used to identify the deficiencies of internal resources such as necessary functionality and knowledge. We examined an extended application of innovation architecture (IA) as a TRM tool that identifies potential partners who complement the necessary resources and then design models of how to cooperate and then explored to launching this application as a social innovation (Tschirky and Trauffler 2008; Igarashi and Okada 2015).

In this chapter, we begin by describing our four practical dementia project cases ("Futures," "Run Tomorrow," "the Fujinomiya Project," and "Words for a Journey") and their mutual collaborations among multiple stakeholders. In these cases, the issues surrounding dementia function to bring together organizations in different sectors by linking boundary objects and inducing external chasm networks. We propose a series of action-development-relationship (ADR) processes as an innovative design methodology for creating strategic partnerships and networks. Our findings demonstrate the usefulness of creating IA components for ambiguous social issues such as dementia.

16.2   Series of Dementia Projects as Weak Coupling Cooperation Cases

Dementia is a social phenomenon that has an enormous impact on society. The estimated ratio of elderly people with dementia in Japan is 15%—a number that exceeds 4.6 million and is projected to grow to 7 million by 2025 (Ministry of Health, Labour and Welfare 2015). On a worldwide scale, the number of people with dementia is expected to increase to approximately 115.4 million by 2050 (World Health Organization 2012).

It is nearly impossible for a single organization or sector to find a clear solution to such a complex and wide-ranging social issue. National and local governments (the public sector) and the medical establishment (the healthcare sector) have implemented helpful dementia-related initiatives, but there are some limitations on what can be achieved by individual sectors. What is clearly needed is a broadly based social innovation involving the participation of both public and private sectors, which is something that by definition goes beyond the conventional framework.

The goal is to develop pan-sector collaboration and also to operate a multiple spiral SECI model of knowledge dimensions (Nonaka and Takeuchi 1995). The SECI model shows the relationship and conversing process between tacit knowledge and explicit knowledge. It is also necessary to clarify how to orchestrate the collaboration within the sphere of ambiguous social issues such as dementia.

Fig. 16.1 (**a**) Run Tomorrow, (**b**) Futures, (**c**) the Fujinomiya Project, (**d**) Words for a Journey

Fujitsu Laboratories Ltd., GLOCOM (Center for Global Communications, International University of Japan), and the nonprofit organization Dementia Friendship Club launched a collaborative project to chart a new way of dealing with dementia-related issues in 2011. In 2013, we jointly set up a networked organization—the Dementia Friendly Japan Initiative—with other partner organizations and sectors. The objectives of the initiative are to have a better understanding of the impact of dementia and to collect and benefit from the wisdom of the various organizations.

Since 2011, we have organized successive projects with actors in the private, public, and community sectors. These projects, shown in Fig. 16.1, are "Run Tomorrow," "Futures," "the Fujinomiya Project," and "Words for a Journey." Each project consisted of a variety of events and activities led by members of different sectors, and each had very distinct characteristics.

Run Tomorrow, which is a long-distance relay involving people living with dementia, brings together people from various sectors of local communities across Japan (Dementia Friendship Club 2011–2015). To realize the goal of a dementia-friendly society, people with dementia, their families, friends, young people, and elderly people each run a short distance, traversing the Japanese islands and passing on a sash to each other one by one. Run Tomorrow delivers the strong message that "dementia or not, should come enjoy life together." It started as a tiny movement where 171 people ran 300 km from Hakodate to Sapporo in Hokkaido. Four years later in 2015, it has expanded into a huge project involving 8000 people, running a total distance of 3000 km from Hokkaido to Kyushu. Run Tomorrow spread out all over Japan.

With Futures, a project jointly developed with the British Council, we brought together entrepreneurs and intrapreneurs from various sectors in the UK and Japan including the business sector, the public sector, social enterprises, community groups, and academia (British Council 2012). Attendees conducted field research and engaged in dialogue under the common theme of an aging society. Futures worked as an incubator of new relationships. The Fujinomiya Project was born by utilizing the efforts of the UK project "Historypin" and a Run Tomorrow promotional film created by a young film director involved with Futures. The most valuable takeaway from Futures is that it created new relationships among stakeholders.

The Fujinomiya Project was an effort to create a trans-generational experience by using old photos of Fujinomiya City, a medium sized located 150 km west of Tokyo and with a population of 130,000 (GLOCOM 2013). In this project, high school students in Fujinomiya set up an interactive event with senior citizens, including individuals with dementia. This interactive event was facilitated by high school students, the shopping district of Fujinomiya, and local community resources. The project strengthened the relationships of the people involved with Run Tomorrow and brought new relationships and contacts to Fujinomiya.

Words for a Journey is the result of collaborative work between Keio University and the Dementia Friendly Japan Initiative (Iba et al. 2015). It is a new pattern language consisting of 40 patterns designed to enable better living with dementia. The patterns were constructed after extensive interviews with people with dementia, their families, and others in their support network. With this endeavor, we combined our experience, knowledge, and the relationships we had developed in previous projects. The patterns created by this project function as a new boundary object combining people and new stakeholders.

Figure 16.2 shows the corporate diagram for each project. Each consists of three themes: "Open Innovative Society" led by the British Council, "Dementia Friendly Community" led by Fujitsu Laboratories and DFC, and "Open Data Society" led by GLOCOM. The resources of the three themes' actors overlap, and internal communication exists between the three. Each project has a mutual relationship, and the members also partially overlap. This resource sharing accelerated the activities and helped increase leverage. In the ambiguous field of social affairs, this approach is the key to success.

However, the relationships between the projects had rather weak couplings, which we believe is also a significant aspect. If we achieved strong coupling between projects, it would necessitate higher levels of capital contribution in terms of human resources and expenditure, which would slow down the social collaborations. To promote flexibility of cooperation, it is necessary to create and strengthen networks of innovators and early adopters.

The series of projects creates a chasm internal network. The conceptual scheme of the chasm internal network applying the chasm diagram is shown in Fig. 16.3 (Moore 1991). The social issues surrounding dementia bring together organizations and other players across sectors, because not only is it an attractive challenge for innovators and early adopters, it also creates bonds of common interest across

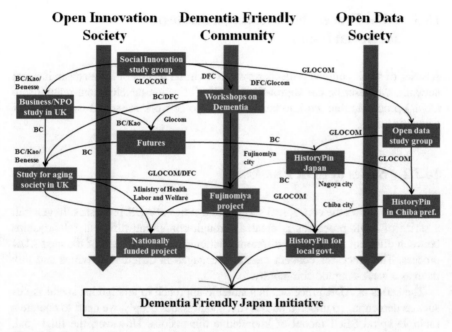

Fig. 16.2 Corporate diagram for each project

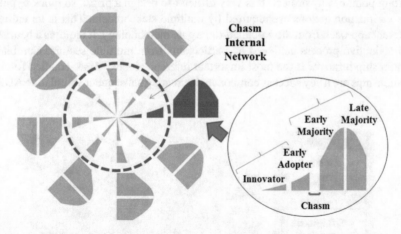

Fig. 16.3 Internal chasm network

society. Social issues create centripetal force and function as boundary objects. The chasm internal network expands the innovators and early adopters to early majorities and accelerates the process of open social innovation. It is essentially similar to the dawn of Internet history. Figures 16.2 and 16.3 show that the issues related to dementia function as boundary objects to combine the people and stakeholders outside.

16.3   ADR Process for Creating a Strategic Network
for Social Issues

A series of weak-coupled projects serves as a virtual incubator for a chasm internal network. We describe our approach as a series of action-development-relationship (ADR) processes that work to create a strategic network for social issues such as dementia issues.

16.3.1   Series of ADR Processes

Figure 16.4 shows the conceptual scheme of a series of ADR processes. In general, a series of ADR processes is created through combining different relationships between different ADRs. Two different relationships are the seed of the next ADR process. The processes connect members engaged in different activities and link them as a weak-coupled community.

The series of ADR processes is a suitable approach to ambiguous social issues such as dementia. To understand relevant social issues deeply, we need to operate a multiple spiral SECI model of knowledge dimensions. However, the final goal, which stakeholders engage, and the required methodology are quite vague at the starting point of any project. It is very difficult to design a priori, so quick hypothesis verification actions are required by multiple stakeholders. This is an entirely different approach from the so-called stage gate methodology. It requires a heuristic and abductive process achieved by participants from multiple sectors. Combined relationship turnover is the most important reference index. Weak-coupled foreign relationships are a key success component in creating subsequent stimulative ADRs.

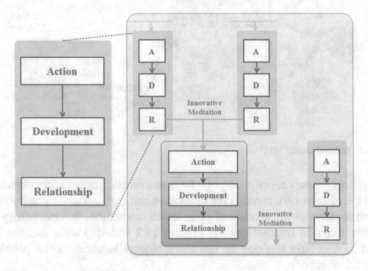

Fig. 16.4 A series of ADR processes

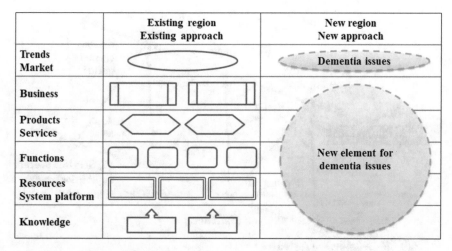

Fig. 16.5   Innovation architecture (IA)

16.3.2   ADR Process in IA Methodology

Innovation architecture (IA) is a technology road-mapping (TRM) tool that visualizes how different types of knowledge and functional elements are interrelated. It is also a very useful communication tool with which to design and evaluate composing elements for new products, services, functions, resources, and knowledge in a new trend or market. Figure 16.5 shows the conceptual scheme of IA.

However, it is not easy to define the necessary elements for a new region or approach in IA. It is even more difficult for ambiguous social issues such as dementia because social issues require contributions from multiple sectors. This means that quick resource procurement is the key to success for innovation.

To identify new elements in each category in Fig. 16.5, ladder actions are often applied. Ladder actions have two layers: a layer of concept and a layer of practice, as shown in Fig. 16.6. In particular, the early phase of the actions is made clearer by on-the-spot observations. Through the ladder actions, the necessary elements of IA are discovered. However, it is not easy to operate the ladder actions simply for innovation processes because not all the elements are clear — specifically, the Stakeholders and their knowledge, resources, available functionalities, and roles have not been clarified. Recursive actions are thus required in the early stages of ladder actions to make the elements in Fig. 16.5 clearer.

The series of ADR processes affords participants from multiple sectors with mutual learning opportunities in these situations. This is a cultivating procedure to make elements progressively clearer. Effective new resources and functionalities in each sector are discovered through connections between members.

In the cases shown in Fig. 16.1, we gradually learn the necessary elements in Fig. 16.5. Through combining two ADRs, we enlarged the opportunities and

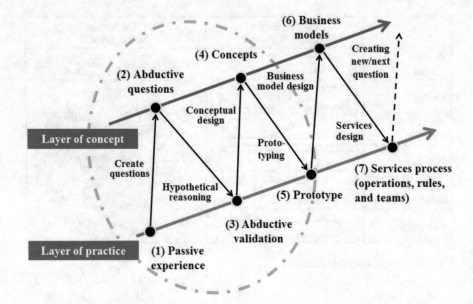

Fig. 16.6 Ladder actions to develop a business model

possibilities by securing resources from outside; that is what happened in our dementia cases. The recursive series of ADR processes created a new set of abductions and made the components of IA clearer. The series of ADRs is equivalent to multi-thread ladder actions. This is very important from the viewpoint of setting new elements to IA. The intrinsic value of the series of ADRs lies not in creating new business in this context; rather, its value is that it makes the IA elements clearer for ambiguous social issues such as dementia.

16.4 Conclusion

We presented our practical series of cases of dementia projects consisting of mutual collaborations between multiple stakeholders. In these cases, the issues surrounding dementia function to bring together organizations in different sectors as linking boundary objects and induce external chasm networks. We also described how our dementia projects work as a series of ADR processes, and why it is useful to create IA components for ambiguous social issues such as dementia.

Acknowledgments This project has been made possible thanks to the support and cooperation of many people. We extend special thanks to everyone involved with "Run Tomorrow," "Futures," "the Fujinomiya Project," "Words for a Journey," and the Dementia Friendly Japan Initiative.

References

British Council (2012) Futures: inspiring social innovation. http://www.britishcouncil.jp/en/programmes/society/futures. Accessed 15 May 2016

Dementia Friendship Club (2011–2015) Run Tomorrow. http://runtomorrow.jp/. Accessed 15 May 2016

GLOCOM (2013) The Fujinomiya Project. https://vimeo.com/110885509. Accessed 15 May 2016

Iba T, Matsumoto A, Kamada A, Tamaki N, Matsumura T, Kaneko T, Okada M (2015) A pattern language for living well with dementia: words for a journey. In: Abstracts of the collaborative innovation networks (COINs 15), Keio University, Tokyo, 12–15 Mar 2015

Igarashi Y, Okada M (2015) Social innovation through a dementia project using innovation architecture. Tech Forcasting Soc Chang 97:193–204

Ministry of Health, Labour and Welfare (2015) Estimated number of the elderly with Dementia. In: Comprehensive strategy to accelerate dementia measures (new orange plan). http://www.cas.go.jp/jp/seisaku/ninchisho_taisaku/dai1/siryou1.pdf. Accessed 15 May 2016

Moore G (1991) Crossing the chasm: marketing and selling high-tech products to mainstream customers. HarperCollins, New York

Nonaka I, Konno N (2012) The grammar of knowledge creating management for prudent capitalism. Toyo Keizai, Tokyo, pp 32–34

Nonaka I, Takeuchi H (1995) The knowledge-creating company: how Japanese companies create the dynamics of innovation. Oxford University Press, New York, p 284

Okada M, Igarashi Y, Nomura T, Tokuda T (2013) The Dementia Project: innovation driven by social challenges. FUJITSU Sci Tech J 49(4):448–454

Star SL, Griesemer JR (1989) Institutional ecology, 'translations' and boundary objects: amateurs and professionals in Berkeley's Museum of Vertebrate Zoology, 1907–39. Soc Stud Sci 19(3):387–420

Tschirky H, Trauffler G (2008) Developing TRM in practice using the innovation architecture. In: A design/induction/enforcement of technical road map and practical use to a research and development strategy. Technical Information Institute, pp 106–136

World Health Organization (2012) Dementia: a public health priority. http://www.who.int/mental_health/publications/dementia_report_2012/en/. Accessed 15 May 2016

Chapter 17
Depression as a Global Challenge and Online Communities as an Alternative Venue to Develop Patients-Led Supportive Network

Sayaka Sugimoto

17.1 Introduction

People suffering from major depressive disorder are increasing worldwide. Today, an estimated 350 million people of all ages in the world suffer from the illness (World Health Organization 2012). It is the leading cause of disability for people of all ages and both genders around the world (World Health Organization 2012). The illness is also prevalent in Canada. Approximately, 2.5 million people are reported to have a depressive disorder (Zoutis 1999). There is a pressing need for a comprehensive, coordinated response to the disorder at country level (World Health Organization 2012).

Effective treatments are available. However, people are reluctant to seek treatments in traditional healthcare settings because of the social stigma associated with mental illnesses. Patients often experience the negative responses of others as well as their own responses to depression while they seek formal treatments (Barney et al. 2006). Even if patients access formal services, they often experience the inadequate levels of services, such as limited access to high-quality primary and non-pharmacological care as well as stigmatizing attitudes of many healthcare professionals (McNair et al. 2002). Moreover, the availability of services at formal healthcare institutions is restricted to their service hours. While the severity of depression tends to increase in the morning, night, or during holidays, hospitals or clinics may not be providing consultation.

Providing support through online communities may be particularly appropriate for people living with depression. Unlike traditional, face-to-face care provision systems, online resources offer additional advantages associated with their power to transcend distance, time, and psychological barriers (Johnsen et al. 2002;

S. Sugimoto (✉)
Juntendo University, Tokyo, Japan
e-mail: sayakasugimoto@gmail.com

© Springer International Publishing Switzerland 2016
M.P. Zylka et al. (eds.), *Designing Networks for Innovation and Improvisation*,
Springer Proceedings in Complexity, DOI 10.1007/978-3-319-42697-6_17

Shaw et al. 2000; Wright 2000). The availability of online services 24 h a day and 7 days a week is a strength lacking in formal healthcare institutions. Moreover, the ease of access is also beneficial for those who have limited mobility due to the severity of the illness. The Internet can, therefore, serve as an alternative place for depression patients to seek information and support. Among many other resources available on the Internet, online support groups have shown strong potential to foster supportive and resourceful environments. An online support group is a group of individuals with similar or common interests who interact and communicate through a computer communication network; this allows social networks to build over a distance (Eysenbach et al. 2004, p. 1). In fact, depression only communities are one of the most popular types of health-related support groups, along with those for breast cancer and HIV patients (Davidson et al. 2000). In depression online support groups, participants can anonymously exchange information and support with others living with the same illness (Muncer et al. 2000).

However, many aspects of depression support groups are still unknown due to the paucity of research. Little is known about who use those groups, why and how those who participate in those groups use online support groups, and what the potential advantages and disadvantages or risks are. Further investigation is necessary to evaluate online support groups in the management of a major depressive disorder.

17.2   Methods

The Depression Centre (http://www.depressioncentre.net) was selected as the data collection site for this study. The Depression Centre is an eHealth program launched in January 2001 by Evolution Health, a private company located in Toronto, Canada. The Depression Centre is one of the eHealth programs and was created under the supervision of a Professor of Psychiatry and Psychology at Ryerson University. It is designed for people living with depression and their caregivers. It consists of multiple interactive components: the Depression Program, the WB-DAT, Session Diary, Symptom Tracker, glossaries, and the Forum. The Forum takes the format of a bulletin board in which users interact asynchronously. The frequency of post is approximately 2–3 posts per day, and the length of posts can vary from 5 to over 100 words.

Mixed methods with a concurrent triangulation strategy were employed to explore the research question. A purposive sampling approach was conducted to select 980 posts made in the Forum. Demographic and clinical information about the users who created those posts was retrieved and recorded. Subsequently, quantitative and qualitative content analyses were conducted to examine what types of support were requested and provided through the posts. A predetermined coding scheme and coding sheets were used to classify and record the types of posts. Posts that sought specific type of support in a question format were coded as explicit, while those posts that implicitly sought support in a non-question format were recorded as implicit. Inter-coder reliability was calculated to ensure consistency of the coding process. The quantitative data about users' support exchange showed the total amount and frequency of each support type, while the qualitative data offered richer information about the nature of the support exchange.

17.3 Results

The sample posts made between January 1 and December 31, 2010 were retrieved from the Evolution Health server on February 24, 2012. Of the 980 posts, 650 posts were created by 67 users and 230 posts were made by 9 mediators, trained employees of Evolution Health. The number of support requests ranged from 1 to 36 per user, while the average number of requests made was 5.

In total, 52 users created 260 posts that contained requests for support and 606 posts that were considered as provisions of support. Table 17.1 (see Appendix) provides the descriptive statistics of the number of instances of support requests and provisions organized by support category. Users exchanged informational support, emotional support, coaching support, companionship, instrumental support, and technical support from other users. There were no instances of requests for spiritual support.

The qualitative analyses of support exchange revealed some overarching themes across categories. The difficulty of managing depressive symptoms was the most frequently discussed topic in all of the instances of explicit requests for encouragement and affirmation, as well as in explicit requests for coaching support in managing psychological symptoms. The fact that users sought both emotional and coping support indicates that users not only want encouragement and sympathy but also to learn how to cope with those symptoms from peers.

Another overarching theme found across different support categories was difficulty managing interpersonal relationships while managing depression. It is the most frequent topic in explicit requests for understanding and coaching and in implicit requests for encouragement, affirmation, and understanding. The qualitative analyses of posts illustrated that users struggled with managing positive relationships with friends, coworkers, boyfriends and girlfriends, marriage partners, family members, and other people in public. A closer examination of posts showed that depression could be both the cause and result of negative relationships. Issues of stigma and exposure were also frequently discussed among users, as these factors block users' efforts to focus on their treatments.

17.4 Discussion

From the quantitative and qualitative exploration of support exchange among users, some aspects of users' lives that generate various support needs began to emerge.

Quantitative analyses of user information suggest that people who come to the Forum have substantial needs for a wide range of additional support regardless of whether they were receiving any formal treatment. Among the 67 users in this study, 66 identified themselves as patients of major depressive disorder and more than half (42) stated that they were receiving formal treatment at the time of registration. Qualitative examination of the posts often discovered discussion among users regarding ongoing struggle with medications, psychotherapy, and healthcare professionals, which further reinforces the idea that users seek additional support from peers while receiving formal treatment.

Qualitative analyses of the posts, in fact, demonstrated how, for many users, depression is not the sole source of distress in their lives. Struggles related to managing depression and work also appeared frequently across categories. Typically, users expressed difficulty with regard to making the decision to take a sick leave from work, suffered from sense of guilt during the break, and experienced mixed feelings of achievement and frustration after going back to work. Moreover, many users mentioned other health problems (experienced by them or their family members), dysfunctional marriage/family relationships, the loss of important other people, loneliness, and financial challenges. The presence of these struggles appeared to be associated with worsening depressive symptoms. Equally, depressive symptoms appeared to reduce the ability of users to manage their life struggles, creating a vicious circle.

This study found substantial unmet information needs among users. The qualitative analyses of requests for informational support revealed the lack of understanding among users about depression and treatments. Users expressed confusion about causes and symptoms of depression, and their lack of understanding often resulted in frustration. In terms of treatments, users were most interested in learning about medications. In fact, about half of the explicit requests for informational support are related to questions about specific medications, such as the long-term effectiveness, side effects, and costs. Examination of posts also revealed some misunderstanding about formal treatments among users. Some users expressed fear of being locked up, and they expressed difficulty in communicating with doctors for this reason. Others expressed their concern about being forced to take a treatment against their will.

The qualitative exploration of the posts generated multiple insight on how similarities and differences between users were reflected in the contents of their conversation. The commonness among the users appeared to construct the foundation of the community. For instance, there were numerous cases in which users expressed their appreciation for the fact that others in the community could truly understand their feelings, could relate to their experience, and could be genuinely sympathetic to their situation due to the common experience of living with the illness. The frequency of the request and provision of understanding, a type of emotional support, demonstrated that users' need to find someone who could relate to their feelings or experience. The fact that the majority of users were receiving formal treatment appeared to be strongly related to their informative discussion over medications and their effectiveness, side effects, and withdrawal symptoms. On the other hand, differences among users appeared to be contributing to the generation of useful advice and creative ideas. When male users were struggling with their relationships with their wives or daughters, for example, several female users with similar experience provided their opinions as a possible interpretation of the situations from the wives or daughters' perspectives. As the results section demonstrated, the diversity among users appeared to be contributing to their creative discussion over unique, nonmedical methods to manage depression. Future studies should further explore the relationships between user characteristics and support exchange to generate more definitive principles.

The qualitative exploration of users' posts provided a deeper insight into the ways depression online support groups functioned in the lives of users. The posts exchanged among the users portray the depression online support group as a place with accepting and supportive atmosphere. It appeared to serve its users as a place to meet sympathetic others with similar experience; place to realize that they were not alone; to discuss what they could not discuss elsewhere; to "just vent" with or without the expectation to have someone to listen.

17.5   Limitations

Content analysis can only be interpreted from the messages left publicly in the Forum. Although this study aimed to explore the roles of depression online support groups in the management of depression described through the posts made in the Forum, how the support exchange through these posts actually changed users' behaviors or attitudes is unknown. This study has also limited ability to assess users' level of satisfaction or dissatisfaction with responses that they received from others in the support group. Although this study aimed to explore the roles of depression online support groups in the management of depression described through the posts made in the Forum, how the support exchange through these posts actually changed users' behaviors or attitudes is unknown.

The present study is exploratory in nature, and thus, more studies need to be conducted to examine to what extent the findings from the present study can be generalized. In particular, the findings from this study have limited generalizability to other depression online support groups without moderators.

17.6   Conclusion

The broad purpose of this study was to contribute to a better understanding of online support groups in the context of major depressive disorder. More specific goals of this study were to examine support exchange among users of a depression online support group and to explore various roles that the support group could play in the management of depression. The findings from this study indicate that unmet emotional, coaching, and informational support needs clearly exist among many users whether or not they are receiving formal treatment. The quantitative analyses of user characteristics, for example, demonstrated that both those who were and were not receiving formal treatment sought emotional support at the Forum. The findings suggest that unmet needs exist because, since depression and other life struggles are so closely intertwined, formal treatment alone cannot completely meet the emotional needs of users. The present study informs healthcare professionals and information professionals of the need for care coordination across different support providers.

Overall, this study indicates that depression online support groups have a strong potential to provide people living with major depressive disorder with valuable support. The study demonstrated that users could seek information, understanding, encouragement, affirmation, and advice to manage physical symptoms, psychological distress, and socioeconomic problems, as well as to find companionship in the Depression Centre Forum. They can seek support any time they want. They can also spend as much time and as many words as they want. For many users, it was important to know if others living with the same illness have gone through similar experience or had similar feelings. Yet, gaining understanding from others is difficult through other sources of information and support. Many users, in fact, expressed their hesitation to disclose their illness to others, including their family members and close friends. In depression online support groups, users can express their emotional needs freely and seek encouragement or sympathy from others without worrying about being judged negatively. It was notable that users not only sought someone who listen to them but also sought advice and suggestions regarding how to cope with the situations that they were going through. It shows users interest and will to learn how to manage their illness better.

Based on the findings of this study, one can hypothesize that users of depression online support groups can gain more skills to adopt and manage their illness through their involvement in the group. In testing this hypothesis, it is important to pay attention to the degree of satisfaction among users and the degree of success in their support exchange. The present study discovered that while some threads ended with an expression of gratitude, other threads ended abruptly without any comment from the person who made the initial posts. This observation indicates that not all users of depression online support groups receive the types of support they wanted in the way that satisfied their needs. Future studies should explore these areas to further examine the potential of online support groups to relieve suffering among people living with depression.

Acknowledgments This research was conducted as a part of my doctoral studies at the University of Toronto. I would like to thank the Evolution Health (Trevor van Mierlo and Rachel Fournier) for providing me with the data necessary to conduct the research and the University of Toronto (Professors Juris Dilevko, Alex Jadad, and Aviv Schachak) for providing me with necessary training and supervision.

Appendix

Table 17.1 Demographic characteristics and clinical status of the users who sought or provided support at the Depression Centre Forum

Support category/sub category		Instances of explicit request of support	Instances of implicit request of support	Instances of support provided (provided by health educator)	Total instances of support
Informational support		29	3	91 (17)	123
Emotional support	Understanding	24	4	109 (7)	137
	Encouragement	24	104	140 (47)	268
	Affirmation	20	69	258 (124)	347
Coaching support	Psychological	25	3	157 (72)	185
	Physical	2	2	10 (5)	14
	Socioeconomical	19	2	55 (25)	76
Companionship	Chat	5	2	70 (5)	77
	Group cohesion	1	0	4 (2)	5
Spiritual support		0	0	3 (0)	3
Instrumental support		0	1	2 (0)	3
Technical support		2	3	4 (4)	9
Total		**151**	**193**	**903**	**1247**

Table 17.2 Descriptive statistics for number of instances of support requests and provisions

		Characteristics of all users	Characteristics of support seeker	Characteristics of support provider
Total number of users		67	52	48
Gender	Number of males	23	18	16
	Number of females	44	33	31
Age	Average (mean)	38.4	38.7	38.9
	Median	35.5	35	37.5
	Mode	51	51	51
Clinical status	Number of users who identified themselves as depression patient	66	51	47
	Depression rate (mean)	6.8	6.9	6.3
	Depression interference (mean)	6.3	6.5	6.1
	Depression level (mean)	6.2	6.4	6.0
	Currently receiving formal treatment	42	33	28
	Experience of CBT	31	22	20

(continued)

Table 17.2 (continued)

		Characteristics of all users	Characteristics of support seeker	Characteristics of support provider
Number of posts users made	Average (mean)	11.2	5	8.3
	Median	4	2	2.5
	Mode	1	1	1
Countries of residence		Canada (31), USA (25), UK (5), Australia (5), India (2), Hong Kong (1), Malaysia (1), South Africa (1), United Arab Emirates (1), Unknown (7)	USA (21), Canada (16), UK (3), Australia (2), Hong Kong (1), India (1), Malaysia (1), South Africa (1), Unknown (5)	Canada (18), USA (17), UK (4), Australia (2), India (2), Hong Kong (1), United Arab Emi (1), Unknown (3)

References

Barney LJ, Griffiths KM, Jorm AF, Christinsen H (2006) Stigma about depression and its impact on help-seeking intentions. Aust N Z J Psychiatry 40(1):51–54

Davidson KP, Pennebaker JW, Dickerson SS (2000) Who talks? The social psychology of illness support groups. Am Psychol 55(2):205–217

Eysenbach G, Powell J, Engelsakis M, Rizo C, Stern A (2004) Health related virtual communities and electronic support groups: a systematic review of the effects of online peer-to-peer interactions. Br Med J 328:1166–1172

Johnsen J, Rosenvinge J, Gammon D (2002) Online group interactions and mental health: an analysis of 3 online discussion forums. Scand J Psychol 43:445–449

McNair BG, Highet NJ, Hickie IB, Davenport TA (2002) Exploring the perspectives of people whose lives have been affected by depression. Med J Aust 176:69–76

Muncer S, Burrows R, Pleace N, Loader B, Nettleton S (2000) Births, deaths, sex and marriage, but very few presents? A case study of social support in cyberspace. Crit Public Health 10(1):1–18

Shaw B, McTavish F, Hawkins R, Gustafson DH, Pingree S (2000) Experiences of women with breast cancer: exchanging social support over the CHESS computer network. J Health Commun 5:135–159

World Health Organization (2012) Depression. Retrieved 10 Mar 2012, from http://www.who.int/mental_health/management/depression/definition/en/

Wright K (2000) Perception of online support providers: an examination of perceived homophily. Commun Q 48(1):44–59

Zoutis P (1999) Ontario mental health statistical sourcebook: volume 1: an investigation into the mental health supplement of the 1990 Ontario health survey. Canadian Mental Health Association, Toronto

Appendix A: Leadership, Communication, and Charisma

Agostino La Bella

Leadership within social, economic, and/or institutional organizations is generally perceived as the ability to influence and guide people without making use of hierarchical power. In spite of the importance given to this concept in the current organizational behavioral thinking, and the enormous literature on the subject, we are far from a universally accepted "leadership model" and its consequent prescriptions to follow in order to attain such a role. In this talk, we will focus on charisma, one of the attributes that all great leaders possess. Literally meaning a "divine gift," the term charisma has been used to indicate the ability to use symbols, emotions, and ideologies to attract and inspire followers. On a wide scale, it often assumes a heroic nature: the commander revered by his soldiers, the head of a fanatical group worshipped by his followers, the politician acclaimed by the cloud, the corporation head that becomes a worldwide symbol of his company's products, the trade union leader that rouses the workers. However, we are interested here in a more concrete and prosaic view of charisma, i.e., the ability of not going unnoticed, to command and maintain attention, to arouse positive emotions, to successfully uphold opinions, ideas, and projects within the comparatively smaller context of our working, social, and family relationships. In this frame, we will show how individual charisma can be built using the several facets of interpersonal communication, focusing in particular on the conscious and subconscious mental processes that determine what is worth of attention. We will also describe possible performance measures for leadership at both the individual and organizational levels.

A. La Bella (✉)
Department of Enterprise Engineering,
Tor Vergata University of Rome, Rome, Italy
e-mail: labella@dii.uniroma2.it

© Springer International Publishing Switzerland 2016   181
M.P. Zylka et al. (eds.), *Designing Networks for Innovation and Improvisation*,
Springer Proceedings in Complexity, DOI 10.1007/978-3-319-42697-6

Appendix B: Building Collective Consciousness—Homo Collaborensis

Peter A. Gloor

In his science fiction stories, Isaac Asimov predicts a future where our successors, bodiless and collectively intelligent, cohabit trillions of worlds in shared consciousness with universal artificial intelligence. While we are far away from this vision, the Internet, Google, and Wikipedia help us for the first time to combine individual realities into one global shared consciousness. This talk will describe, how, through the application of social quantum physics, we will become more collaborative, and what steps we can take today towards becoming "homo collaborensis." "Honest signals of collaboration" consisting of words and visual cues will lead through better communication to better collaboration, resulting in more innovation. These "honest signals" will create entanglement between team members, who will—through understanding and visualizing these social cues—change their behavior towards better collaboration. In this sense, their behavior of today will change their interpretation of the past, leading to a change in their future behavior. The six honest signals of collaboration are strong leadership, balanced contribution, rotating leadership, responsiveness, honest sentiment, and shared context. They will be described in detail and illustrated with many practical examples. The talk will lay out what these signals mean for each of us, and what we can do to become better communicators, leading to becoming better collaborators, and thus ultimately become more creative, leading more fulfilling lives.

P.A. Gloor (✉)
MIT Center for Collective Intelligence,
Cambridge, MA, USA
e-mail: pgloor@mit.edu

© Springer International Publishing Switzerland 2016   183
M.P. Zylka et al. (eds.), *Designing Networks for Innovation and Improvisation*,
Springer Proceedings in Complexity, DOI 10.1007/978-3-319-42697-6

Appendix C: Words and Networks: Using Natural Language Processing to Enhance Graphs and Test Network Theories

Jana Diesner

The structure and dynamics of collaboration and communication networks are impacted by both social interactions and information exchange between people. In this talk, I present our research on using natural language processing techniques to enhance social network data. The ultimate goal with this work is to test the validity of classic social network theories in today's contexts. I show our findings from leveraging sentiment analysis to label edges in communication networks in order to enable triadic balance assessment. In another example, we studied the homogeneity or diversity of clusters in networks with respect to basic principles of morality. I address methodological challenges, such as the validation and adaptation of lexical resources, and the directed nature of ties in communication networks. The presented methods enable the scalable and systematic detection of edge properties, which reduces the need for surveys or manual link labeling. Finally, I briefly discuss challenges in working with "open data" from online sources.

J. Diesner (✉)
The iSchool (Graduate School of Library and Information Science),
University of Illinois at Urbana-Champaign,
Champaign, IL, USA
e-mail: jdiesner@illinois.edu

© Springer International Publishing Switzerland 2016
M.P. Zylka et al. (eds.), *Designing Networks for Innovation and Improvisation*,
Springer Proceedings in Complexity, DOI 10.1007/978-3-319-42697-6

185

Printed in the United States
By Bookmasters